Ergogenesis Indexation and Pre-Eumenorrheic Malady in Athenaeum Gynic

Sapientia P.

Ergogenesis Indexation and Pre-Eumenorrheic Malady in Athenaeum Gynic

First Edition December 2025

Written by Sapientia P.

TABLE OF CONTENTS

Sl. No.	Title	Page No.

LIST OF TABLES

LIST OF FIGURES

Chapter I

INTRODUCTION

1.1 Introduction

The slogan "Healthy Women, Healthy World" reflects the fact that, as custodians of family health, women play a critical role in maintaining the well-being of their communities (Blumel et al., 2000). Physical activity encompasses all types of body movement, from the simplest to the most complex. It may be voluntary, including structured and planned activities aimed at improving specific aspects of physical fitness or energy expenditure, or it may be part of daily life activities, such as walking, household chores, occupational tasks, or transportation.

Physical activity can also be spontaneous and involuntary, ranging from minor movements like blinking to full-body muscle contractions associated with posture and mobility.

However, it is often difficult to separate and quantify these different domains, which is why they are frequently considered together (Teixeira et al., 2006). Regular physical activity has been shown to reduce the risk of premature death and prevent the development of numerous chronic diseases (Warburton, Nicol, & Bredin, 2006).

It is currently recommended by the Chief Medical Officer for England that adults take some moderate or vigorous physical activity every day and a minimum of 150 minutes per week (Davies, S. C., Fowler, T., Watson, J., Livermore, D. M., & Walker, D. (2013). However, these minimum recommended levels of physical activity are met by only 39% of men and 29% of women in England. (Watts, P., Phillips, G., Petticrew, M., Hayes, R., Bottomley, C., Yu, G., ... & Lock, K. (2013).

The propagation of the species is a basic aim of nature. In the multiplication of the human race, "woman has a pivotal role to play." Menopause is generally defined as cessation of periods for 12 months or a period equivalent to 3 previous cycles or as time of cessation of ovarian function resulting in permanent amenorrhea. (Modi, M. B., Donga, S. B., & Dei, L. (2012). The menopause is thus a gradual and natural transitional phase of adjustment between the active and inactive ovarian function and occupies several years of a women's life and involves biological and psychological changes adjustments. This period is usually associated with unavoidable manifestation of aging process in women. (Mashiloane, C. D., Bagratee, J., & Moodley, J. (2002). Most women experience near complete loss of production of estrogen by their mid-fifties. (Menopause Guidelines Revision Task Force, A. A. C. E. (2006). During reproductive years, women are protected by female hormones, i.e. estrogen and progesterone. With menopause, women enter an estrogen deficient phase in their lives, which accelerates the ageing process resulting into greater vulnerability to psychosomatic problems. Hot flushes, sweating, changes in mood and libido are some of the important outcomes affecting the quality of life (QoL) during climacterium in women. QoL covers physical, functional, emotional, social, and cognitive variables up to 85% of menopausal women. (Blumel, J. E., Castelo-Branco, C., Binfa, L., Gramegna, G., Tacla, X., Aracena, B., ... & Sanjuan, A. (2000).

Enhancing the quality of growth is an important objective of the development paradigm in many developing countries. Better health, education, equal and wider job opportunities to all, trustworthy and transparent people's intuitions, sustainable and cleaner environment, dignity, self-esteem and life security, among others, are key manifestations of the quality of growth. (Jha, S., & Bawa, K. S. (2006).

If the quality of human capital is not good, physical capital and natural resources cannot be properly utilised and growth neither be sustained nor be qualitative. Health is major segment of human capital (Mayer, D. (2000, September).

According to WHO, health is a state of complete physical, mental and social well-being and not merely the absence of disease or infirmity. The health status is usually measured in terms of life expectancy at birth, infant mortality rate, fertility rate, crude birth rate and crude death rate. These indicators of health are determined by numerous factors such as per capita income, nutrition, housing, sanitation, safe drinking water, social infrastructure, health and medical care services provided by government, geographical climate, employment status, incidence of poverty and the like, (Reddy, K. N., & Selvaraju, V. (1994).

Since the industrial revolution, the development of new technologies has enabled people to reduce the amount of physical labour needed to accomplish many tasks in their daily lives. As the availability of new devices has continued to increase, the effects on physical labour and human energy expenditure have grown to include many aspects of the lives of more and more people. The effects of some of these technologies on physical activity are obvious (eg, steam, gas, and electric engines;

trains; cars; and trucks), whereas others are more subtle and complex (e.g., televisions, computers, electronic entertainment, the internet, and wireless communication devices).The use of many of these technologies has been driven by the goal of increased individual worker productivity and reduced physical hardships and disabilities caused by jobs entailing continuous heavy labour. However, the human body has evolved in such a way that most of its systems (e.g., skeletal, muscle, metabolic, and cardiovascular) do not develop and function in an optimum way unless stimulated by frequent physical activity. (Booth, F. W., Laye, M. J., Lees, S. J., Rector, R. S., & Thyfault, J. P. (2008).

Although the technological revolution has been of great benefit to many populations throughout the world, it has come at a major cost in terms of the contribution of physical inactivity to the worldwide epidemic of non communicable diseases. Global, N. C. D., & World Health Organization. (2011). Global Status Report on Noncommunicable Diseases 2010. *Geneva: World Health Organization.*

In 2009, physical inactivity was identified as the fourth leading risk factor for non communicable diseases and accounted for more than (Global, N. C. D., & World Health Organization (2011). Million preventable deaths.3 Comparisons of patterns of participation in physical activity between countries and regions were unachievable until a decade ago, (Bull, F. C., Armstrong, T. P., DixON, T., Ham, S., NEiMAN, A. N. D. R. E. A., & Pratt, M. (2004). Largely due to the absence of stand arise instruments suitable for international use. This barrier caused a so-called collective blind spot, because the evidence for the health benefits of physical activity had grown stronger since the 1970s, (Haskell, W. L., Lee, I. M., Pate, R. R., Powell, K. E., Blair, S. N., Franklin, B. A., ... & Bauman, A. (2007). Underpinning the importance of surveillance data to guide national action. (Bull, F. C., & Bauman, A. E. (2011).

Without a suitable instrument, early efforts to characterize patterns of physical activity frequently used only measures of occupational classification or only estimations of energy expenditure during leisure-time physical activity as the best available indicators, such as in early epidemiological studies (Morris, J. N., Heady, J. A., Raffle, P. A. B., Roberts, C. G., & Parks, J. W. (1953).

The habits and the way of life obtained in period of youth may substantially affect human health, development and quality of life in the future. Young people are often considered to be healthiest population, but, as shows studies, this generation's health is at risk. Changes in the dynamic society also change the values and daily habits. Modern technologies more and more facilitate human lives, but it also reduces the physical activity. Sufficient physical activity (PA) is an essential prerequisite for health maintenance and improvement. Physical activities strengthens musculoskeletal system, reduce risk of the heart and vascular diseases, arthritis and multiple cancer risks, and also have a positive impact on human mental health. Physical activity is „any bodily movement produced by skeletal muscles that result in energy expenditure" (Caspersen, C. J., Powell, K. E., & Christenson, G. M. (1985). Physical activity is closely linked to physical fitness. Physical fitness is „a set of morphological and functional characteristics of the individual, which provides the ability to carry out physical activity" (Caspersen, C. J., Powell, K. E., & Christenson, G. M. (1985). According to studies minimum required health enhancing PA is not less than 150 minutes of moderate intensity or 75minutes of high-intensity activity per week. (US Department of Health and Human Services. (2008).

All activities of daily living, which are made with at least medium intensity and continue at least for 10 minutes without interruption, are suitable in order to maintain health. However, despite the obvious physical and mental health benefits,

49% of Latvia population does not engage in any sport or physical activities and as the main reason they mentioned “lack of time” and ”lack of necessity to be physical active” (Bula-Biteniece, I., & Grants, J. (2010).

Also cognitive factors affect one’s perceptions that there are no benefits from moderate activity, physical activity is not enjoyable, and lack of confidence in one’s ability to be active. Starting studies at university, young people significantly change their learning process organization and everyday activity habits changes too. The approaches of organizing sport activities at the Latvian universities are different from primary education. Physical activities at the universities are mainly organized as participation in team sport trainings or attending optional sport courses. Only in some particular universities study course „Sport” is in compulsory section. Therefore clarification of students’ motivation has essential role in their involvement in physical activities. There are many studies that have investigated the mechanisms of motivation to be physical active (Rodgers, W. M., Hall, C. R., Duncan, L. R., Pearson, E., & Milne, M. I. (2010). Scientific literature describes different motivation theories, which the main conclusion is that a person's internal processes have the determinative role on the personality action.

Since the beginning of time, survival has been a daily struggle. Simply staying alive was physically demanding. Yet, humans in industrialized societies over the last 150 years have become spoiled. We live in our minds so much of the time that we have almost forgotten that we have a body. Many people work in offices who make their living by reading, writing, speaking, and thinking, but seldom by physical labor. Regular physical exercise does the body good. (Curl, W. W. (2000).

Cousins, S. O. B., & Tan, M. (2003). If you have a natural health attitude taking the car to cross the road is out, and a workout in the gym three or four times a week is in. Our bodies demand to be used. Failure to use them results in your muscles deteriorating, at a steady and progressive rate. Thus, everyone should develop good, sensible habits of exercise that they can stick to for the rest of their lives. (Das, A. K., & Subramaniam, S. R. M. P. (2014). The objective of exercise is to improve your physical fitness. Thus, if what you are doing does not improve either your cardiovascular fitness or your muscular strength, then it should not be considered as physical exercise.

1.2 An Active Lifestyle

Historically, except for the last 150 years, the demands of daily life gave everybody all the physical exercise that they really needed to get. People used to walk constantly, and lift and carry burdens. They used to walk to gather food, to work, to the market, and they walked to do virtually everything. Instead of dreaming up ways to work as little as possible, why NOT find ways to make many of your ordinary activities more vigorous? Since man was meant to walk, the easiest way to increase your overall physical fitness is to choose to walk at every opportunity possible. Walking is an ideal form of physical exercise that physically fit individuals can routinely participate in. (Das, A. K., & Subramaniam, S. R. M. P. (2014). Walking, also, provides for direct contact with the elements of nature. Walking tones muscles, improves your fitness, circulation, appetite, and breathing. Furthermore, walking, unlike bicycling or swimming, is a load-bearing exercise. (Curl, W. W. (2000). Choose to walk instead of driving short distances. Choose to take the stairs instead of using the elevator. Try doing yard work, working on your car, and cleaning your house. Participating regularly in some type of physical recreation is even better.

Most girls drop their rate of physical activity once they reach adolescence. Physical activity is an important part of health and wellbeing, and girls should remain active as they grow up. Parents can help their daughters to become more active and healthy. Physical activity is important for everybody, including all teenagers, but especially for girls who are generally less active than boys the same age. As teens juggle the transition from primary school to high school there are other pressures that come with itsocially, at home, and at school. Issues such as body image, the onset of menstruation, and general feelings of insecurity about the changing body can surface in these years. Physical activity sometimes takes a back seat to other priorities. However, physical activity is an important part of health and wellbeing, and girls should remain active as they grow up. Teenage girls should do at least 60 minutes of physical activity every day. This can be built up throughout the day with a variety of activities. Being physically active doesn't mean you have to do competitive sport, or go to the gym – any physical activity is good for you. To reap the greatest benefits, physical activity should be done at a moderate to vigorous intensity.

In addition to building physical activity into your day, also think about reducing the amount of 'sedentary' activities you do (things done lying or sitting down). Limiting the amount of television you watch or computer activity to less than two hours each day may be helpful. (Spear, B. A., Barlow, S. E., Ervin, C., Ludwig, D. S., Saelens, B. E., Schetzina, K. E., & Taveras, E. M. (2007). Regular physical activity is vital for good physical and mental health. It helps improve your overall health and fitness, maintain a healthy weight, reduce your risk for many chronic diseases and promote good mental health.

The National Physical Activity Guidelines for Australians recommend that at least 30 minutes of moderate-intensity physical activity on most, preferably all, days is required for good health. This is the same for women and men. However, only 54 per cent of Australian women meet these guidelines. Some of the barriers women face includes caring responsibilities, body image and perceptions of safety. Physical activity is an important component to a healthy lifestyle. Not only does it make a person feel better physically and mentally, it reduces the risk of developing chronic diseases, such as cancer, high blood pressure, stroke, Type 2 diabetes, coronary heart disease and osteoporosis. It has also been suggested that up to one-half of the functional decline between the ages of 30 and 70 is not due to aging per se, but to physical inactivity.

Current guidelines for physical activity state that children between the ages of five and eleven and youth between the ages of 12 to 17 should accumulate at least 60 minutes of moderate to vigorous intensity physical activity per day. Adults and seniors should accumulate at least 150 minutes of moderate to vigorous intensity aerobic physical activity per week, in episodes of ten minutes or more (Tremblay, M. S., Warburton, D. E., Janssen, I., Paterson, D. H., Latimer, A. E., Rhodes, R. E., ... & Murumets, K. (2011).

Physical inactivity is ranked fourth among the ten leading risk factors for NCDs and accounted for 3.2 million deaths in the world in 2004. Increasing levels of physical activity in populations is seen as crucial in reducing the risk of NCD and is a key strategy in the WHO Global Strategy on Diet, Physical Activity and Health

(DPAS) (World Health Organization. (2006). The items outlined in the resolution are strategies aimed at reducing the risk of chronic non-communicable diseases across populations by addressing "energy in" (diet) and "energy out" (physical activity), through comprehensive, multi-sectoral interventions. The overall goal is to reduce disease and death rates related to unhealthy diet and physical inactivity by facilitating the development of an enabling environment for sustainable changes at individual, community, national, and global levels. The ecological model of healthy and active community living presented in the figure below illustrates the complex layers of influence and multiple levers that need to be addressed in order to impact physical activity in today's complex societal structure. llustration of the ecological model influencing physical activity and activity living (Edwards, P., & Tsouros, A. D. (2006).

A common misperception of health and healthcare in our society is that it relates solely to the treatment of disease. While access to health services is a key determinant of health, the social determinants of health have come to be commonly understood as being some of the most powerful influences on health in modern societies. (Wilkinson, R. G., & Marmot, M. (2003). Health is remarkably sensitive to both the physical and social environments in which we live, work and play. These environments can in turn drive health disparities between people within our communities by promoting beneficial or detrimental health experiences. In many cases, poor health, related to the social determinants of health, is preventable. Treating the root cause may be more effective, in terms of economic and societal costs, than treating the symptoms alone. An individual's health is determined by many things that we encounter in our day-to-day lives. The Public Health Agency of Canada has identified 12 key determinants of health. (WHO Commission on Social Determinants of Health, & World Health Organization. (2008).

It is, de facto, the quality of human health upon which the realization of life goals and objectives of a persona, the community or nation as whole depends. Health is multi-dimensional phenomenon. It is both an end and means of development strategy. The relationship between health and development is mutually reinforcing- while health contributes to economic development, economic development, in turn, tends to improve the health status of the population in a country. Health is also an important entitlement that enhances "capabilities" of the poor people leading to increase in "commodities" and further improvement in health status, (Bloom, D. E., Canning, D., & Sevilla, J. (2004).

Premenstrual syndrome (PMS), also called premenstrual tension (PMT) is a collection of emotional symptoms, with or without physical symptoms, related to a woman's menstrual cycle. While most women of child-bearing age (up to 85%) report having experienced physical symptoms related to normal ovulatory function, such as bloating or breast tenderness, medical definitions of PMS are limited to a consistent pattern of emotional and physical symptoms occurring only during the luteal phase of the menstrual cycle that are of "sufficient severity to interfere with some aspects of life". (Dickerson, L. M., Mazyck, P. J., & Hunter, M. H. (2003). In particular, emotional symptoms must be present consistently to diagnose PMS. The specific emotional and physical symptoms attributable to PMS vary from woman to woman, but each individual woman's pattern of symptoms is predictable, occurs consistently during the ten days prior to menses, and vanishes either shortly before or shortly after the start of menstrual flow. Two to ten percent of women have significant premenstrual symptoms that are separate from the normal discomfort associated with menstruation in healthy women. (Matlin, M. (2011).

Influences on young peoples' physical activity are multi-factorial. (Sallis, J. F., Taylor, W. C., Dowda, M., Freedson, P. S., & Pate, R. R. (2002). A variety of psychological, social and physical environmental correlates of physical activity for young people have been identified. (Van der Horst, K., Paw, M. J. C. A., Twisk, J. W., & Van Mechelen, W. (2007). However, it has proven difficult to apply this information to promote behavior change in population-based research. To develop more effective physical activity interventions it is important to determine the most salient and influential mediating variables in the target population. (Welk, G. J., Wood, K., & Morss, G. (2003).

Youth physical activity behavior (from here on, abbreviated as PA) can be best described as a profile rather than a single entity or indicator. From this perspective, an individual activity profile would be characterized as a complex matrix of behaviors that take place in a range of social contexts, each with its own set of physiological, psychological and sociological determinants and outcomes. The wide range of correlates supports the application of ecological models of behavior to improve understanding of the influences on youth physical activity. (Sallis, J. F., Taylor, W. C., Dowda, M., Freedson, P. S., & Pate, R. R. (2002).

Structural interventions, involving changes to the physical, political and economic environment are a particularly important component in promoting physical activity as they help to address the facilitators and barriers to individual behavior change. For example, if health education messages are aimed at encouraging the population to walk and cycle more but the infrastructure for cycling and walking do not exist – if road safety is poor, pavements, sidewalks and cycle lanes are not provided or people do not feel secure in their communities - then health education strategies will have little impact on the community. The majority of the evidence regarding the effectiveness of physical activity in impacting NCDs has been summarized by the United States Center for Disease Control and Prevention (Davies, S. C., Fowler, T., Watson, J., Livermore, D. M., & Walker, D. (2013). and the United Kingdom National Institute for Health and Clinical Excellence. (Champion, D. J., Ransom, S. M., Lazarus, P., Camilo, F., Bassa, C., Kaspi, V. M., ... & Stappers, B. W. (2008).

Sedentary behavior has been identified as one of the leading preventable causes of death, (Mokdad, A. H., Marks, J. S., Stroup, D. F., & Gerberding, J. L. (2004). and an inverse linear relationship exists between volume of physical activity behavior and all-cause mortality. (Lee, I. M., & Skerrett, P. J. (2001). Moreover, participation in regular physical activity decreases the risk of cardiovasculardisease, (Penedo, F. J., & Dahn, J. R. (2005). Type 2 diabetes,mellitus, osteoporosis, (Malmberg, J. J., Miilunpalo, S. I., Vuori, I. M., Pasanen, M. E., Oja, P., & Haapanen-Niemi, N. A. (2002). depression,(Brosse, A. L., Sheets, E. S., Lett, H. S., & Blumenthal, J. A. (2002). obesity,(Wing, R. R., & Hill, J. O. (2001). breast cancer,(Breslow, R. A., Ballard-Barbash, R., Munoz, K., & Graubard, B. I. (2001). colon cancer, (Slattery, M. L., & Potter, J. D. (2002). and falls in older adults. (Wing, R. R., & Hill, J. O. (2001). Given the numerous health benefits of physical activity participation, various public health guidelines have been published on the recommended volume and intensity of physical activity for healthy adults. The American Heart Association, the US Surgeon General, the Centers for Disease

Control and Prevention (CDC), and the American College of Sports Medicine recommend at least 30 minutes per day of at least moderate-intensity physical activity on most, and preferably all, days of the week. (Thompson, P. D., Buchner, D., Piña, I. L., Balady, G. J., Williams, M. A., Marcus, B. H, & Fletcher, G. F. (2003). Similar guidelines have been adopted for children, (National Center for Health Statistics (US), & National Center for Health Services Research. (2002). although other consensus panels have recommended one

(Marcus, B. H., Williams, D. M., Dubbert, P. M., Sallis, J. F., King, A. C., Yancey, A. K., ... & Claytor, R. P. (2006). or more (Tudor-Locke, C., Pangrazi, R. P., Corbin, C. B., Rutherford, W. J., Vincent, S. D., Raustorp, A., ... & Cuddihy, T. F. (2004). hours of physical activity per day for children. The CDC, American College of Sports Medicine, and Surgeon General further state that physical activity may be incorporated into one's everyday lifestyle and that the daily physical activity requirements may be accumulated over the course of the day in short bouts of 10 to 15 minutes. Finally, the US Department of Agriculture has recommended 30 minutes of physical activity per day to prevent chronic disease and at least 60 minutes per day to manage weight. (Pimentel, D., Zuniga, R., & Morrison, D. (2005).

Despite the numerous benefits of physical activity and the recent attention to specific guidelines, only 32% of US adults and 66% of children and adolescents (based on Healthy People 2010 guidelines) engage in regular leisure-time physical activity. (Barnes, P. M., & Schoenborn, C. A. (2003). Given the many benefits of physical activity and the low prevalence rates, it is imperative that interventions be designed that effectively promote the adoption and maintenance of active lifestyles in large numbers of people. (Etats-Unis. Department of health, human services, Centers for disease control, prevention (Etats-Unis), & National center for health statistics

(Etats-Unis). (2013). Exhaustive reviews of physical activity intervention studies have been conducted that provided quantitative indices of physical activity intervention efficacy, (Foster, C., Hillsdon, M., Thorogood, M., Kaur, A., & Wedatilake, T. (2005). as well as recommendations for efficacious and cost-effective physical activity promotion strategies. (Kahn, E. B., Ramsey, L. T., Brownson, R. C., Heath, G. W., Howze, E. H., Powell, K. E., ... & Corso, P. (2002). However, these previous exhaustive reviews have not addressed issues specific to population subgroups or intervention-delivery modalities. Numerous additional reviews of physical activity interventions have focused on these particular subareas of physical activity intervention research (eg, Blamey and Mutrie) (Blamey, A., & Mutrie, N. (2004). however, these reviews are published in a wide array of journals and formats and often do not address broader intervention issues, such as integration of intervention modalities.

In addition, we believe it is crucial to continue to develop more effective approaches to physical activity promotion. Thus, the purpose of the present review is 2-fold. First, we provide an overview of existing physical activity intervention research, focusing on subpopulations and intervention modalities. Our reviews within each area are not exhaustive or quantitative, because each area has been reviewed in more depth in numerous other reports. Instead, our goal is to provide a single document that provides a qualitative overview of intervention research that emphasizes selected topics of particular importance for improving the population-wide impact of interventions. Therefore, in synthesizing this vast literature, we begin with existing reviews of physical activity research in each area and incorporate in our discussions recent reports of well-designed individual physical activity intervention studies that expand the existing research base or target new areas of research. Our second purpose is to offer new ideas and recommendations to improve the state of the science within each area and, where possible, to propose ideas to help bridge across these existing categories of research.

Specifically, we more closely examine the state of physical activity interventions research within (1) specific populations, such as among different age groups and within underserved populations, and (2) across different delivery modalities, including healthcare or physician-delivered interventions, worksite interventions, mediated interventions, environmental interventions, and multiple behavior change interventions that include physical activity. These categories of intervention research are not mutually exclusive or exhaustive, but we believe improvements in these areas hold promise for improving the population impact of interventions. We then discuss cross-cutting issues that impact multiple areas of physical activity intervention research, including (1) maintenance of physical activity behavior change, (2) theory testing and development, and (3) diffusion and policy implications. Throughout the review, we discuss research that focuses on increasing physical activity behavior within healthy child, adult, and older adult populations. Discussion of the physiological benefits of physical activity behavior among healthy and clinical populations can be found elsewhere. (Marcus, B. H., Williams, D. M., Dubbert, P. M., Sallis, J. F., King, A. C., Yancey, A. K., ... & Claytor, R. P. (2006).

Environmental interventions impact physical aspects of the environment in an attempt to promote physical activity for leisure and transportation purposes. Despite numerous studies examining associations between environmental variables and physical activity behavior, (National Research Council (US). Committee on Physical Activity, Land Use, & Institute of Medicine (US). (2005). Transportation Research Board. very few controlled intervention studies have been conducted. A review (

Foster, C., Hillsdon, M., & Thorogood, M. (2004). of environmental interventions identified 3 multi component environmental interventions conducted in workplace or military settings that used strategies such as providing additional exercise facilities and providing more time and incentives to use these facilities. These interventions showed small increases in rates of physical activity compared with controls, but each study had multiple potential sources of bias, including use of quasi-experimental designs. The authors also identified 19 studies that tested the effects of prompts to use stairs on subsequent use of stairs instead of elevators or escalators. Again, most studies suffered from multiple design flaws and lack of adequate controls, but in general, they revealed weak, short-term effects. The Task Force on Community Preventive Services (Kahn, E. B., Ramsey, L. T., Brownson, R. C., Heath, G. W., Howze, E. H., Powell, K. E., ... & Corso, P. (2002). located 10 studies of sufficient quality to be included in an evidence-based review of environmental interventions to increase access to physical activity. The median estimates from these studies suggest that creating or improving access to places for physical activity can result in a 25% increase in the number of people who are active at least 3 times per week. The Task Force strongly recommends community-level interventions that create or enhance access to places for physical activity. Finally, a recent study that examined change in physical activity among residents living near a newly constructed walking trail found no increases in several indices of physical activity. In summary, although research on the relationship between environmental variables and physical activity behavior continues to grow, the few intervention studies conducted have shown weak effects at best, and these studies have multiple methodological weaknesses

Physical activity occurs during work, transport, domestic, and leisure-time activities. Too little physical activity increases the risks of ischaemic heart disease, stroke, colon cancer, breast cancer, and type 2 diabetes (Molnar, B. E., Gortmaker, S. L., Bull, F. C., & Buka, S. L. (2004). The World Health Organization recommends at least 30 minutes of regular, moderate-intensity physical activity on most days to reduce the risk of disease and injury (WHO, U. (2004).

Lack of physical activity is a problem in many developed countries, and a growing concern for developing countries adopting a progressively "Westernised" lifestyle (Canadell, J. G., Le Quéré, C., Raupach, M. R., Field, C. B., Buitenhuis, E. T., Ciais, P., ... & Marland, G. (2007). Australia is no exception, with only 44% of men and 36% of women achieving sufficient physical activity for health (Merom, D., Phongsavan, P., Chey, T., & Bauman, A. (2006). This inactivity contributes 7% of Australia's disease burden and 10% of all deaths, mostly due to cardiovascular disease and diabetes (Begg, S. J., Vos, T., Barker, B., Stanley, L., & Lopez, A. D. (2008). It also places a substantial burden on the Australian economy through the costs of treatment for physical activity–related disease and injury, lost productivity, and diminished quality of life (Stephenson, J., Bauman, A., Armstrong, T., Smith, B. J., & Bellew, B. (2000).

Interventions to promote physical activity (referred to herein as "physical activity interventions") typically involve teaching individuals the skills to change physical activity behaviour, providing the population with knowledge about physical activity goals or opportunities to be active, or creating a more physically active environment (Kahn, E. B., Ramsey, L. T., Brownson, R. C., Heath, G. W., Howze, E. H., Powell, K. E., ... & Corso, P. (2002). Currently, in Australia, general practitioners (GPs) are relied upon to deliver physical activity interventions (when time permits). In addition, governments provide some encouragement to change physical activity behaviour through local mass media and transport campaigns, but investment is minimal. It is likely that a combination of intervention approaches will be needed to achieve a meaningful change in population participation in physical activity (Bellew, B., Schöeppe, S., Bull, F. C., & Bauman, A. (2008).

Efficient allocation of health resources to different intervention programs hinges on identifying those interventions (or combinations of interventions) that achieve maximum population health benefits relative to cost. However, there is a need for more cost-effectiveness analyses using standardised methods to enable comparison of different types of physical activity interventions (Cobiac, L. J., Vos, T., & Barendregt, J. J. (2009). Although there have been a number of cost-effectiveness analyses of individual interventions (Cobiac, L. J., Vos, T., & Barendregt, J. J. (2009). including two studies reporting cost-effectiveness of GP prescription in the Australian context (Sims, J., Huang, N., Pietsch, J., & Naccarella, L. (2004).

(Dalziel, K., Segal, L., & Elley, C. R. (2006). there has been only one cost-effectiveness analysis, by the National Institute of Clinical Excellence (NICE) in the United Kingdom (Cobiac, L. J., Vos, T., & Barendregt, J. J. (2009). that used a standardised approach to compare multiple interventions. All ten interventions, which involved brief advice by a health professional or referral to an exercise physiologist, were found to be dominant (i.e., cost-saving) when compared with usual care. However, the NICE study focused only on intervention alternatives rather than evaluating a range of complementary interventions (including broader public health approaches incorporating media, active transport, etc.) and the potential benefits of combining interventions as a package.

In Australia, a standardised approach to assessing cost-effectiveness (ACE) has been developed for evaluating interventions in the Australian health care context (Carter, R., Vos, T., Moodie, M., Haby, M., Magnus, A., & Mihalopoulos, C. (2008). These methods are being used to evaluate 150 interventions, focusing on prevention of non communicable disease, including six interventions to promote physical activity. In this paper, we present the ACE results for the physical activity interventions, which range from individualised counselling interventions to broad population health approaches. The cost-effectiveness of each intervention is compared with current practice for physical activity intervention in the Australian population, and from this we derive an optimal intervention pathway for improving population health.

Physical exercise programs have been developed with the aim of improving a range of outcomes, including physical performance, body composition, hemoglobin concentration, immune function, fatigue, psychological well-being and health-related quality-of-life (HRQOL) (Buffart, L. M., Thong, M. S., Schep, G., Chinapaw, M. J., Brug, J., & Van De Poll-franse, L. V. (2012). There is increasing evidence that cancer patients (e.g., breast, colon, and prostate cancer, hematological malignancies) (Bergenthal, N., Will, A., Streckmann, F., Wolkewitz, K. D., Monsef, I., Engert, A., ... & Skoetz, N. (2014). may benefit from physical exercise programs in terms of improvement in fitness levels, physical activity and HRQOL (Solomon, C. G., Hu, F. B., Dunaif, A., Rich-Edwards, J. E., Stampfer, M. J., Willett, W. C., ... & Manson, J. E. (2002).

Walking is a major component of daily physical activity, and is the most common form of exercise (McDonald, C. M., Widman, L., Abresch, R. T., Walsh, S. A., & Walsh, D. D. (2005). Walking is self regulated in intensity, duration and frequency, and can be an important indicator of a person's health and fitness status (Tudor-Locke, C., & Bassett, D. R. (2004). Among healthy individuals, 10,000 steps daily have been estimated to be of value in maintaining desired health benefits (Brown, J. K., Byers, T., Doyle, C., Courneya, K. S., Demark- Wahnefried, W., Kushi, L. H., ... & Eldridge, B. (2003). For cancer survivors, however, it is unclear whether an increase in walking activity results in enhanced functional health. Improved functional status is a primary goal in the rehabilitation of cancer survivors (Pearson, O. R., Busse, M. E., Van Deursen, R. W. M., & Wiles, C. M. (2004). Relatively new techniques allowing unobtrusive long-term activity monitoring (e.g., with the use of pedometers or microprocessor-based accelerometer recorders) may provide a clearer indication of how much an individual actually walks in daily life (Lubans, D. R., Morgan, P. J., & Tudor-Locke, C. (2009). Two reports indicate that the use of activity monitors may significantly increase physical activity levels among children and adolescents (Bravata, D. M., Smith-Spangler, C., Sundaram, V., Gienger, A. L., Lin, N., Lewis, R., ... & Sirard, J. R. (2007). as well as in adults with and without chronic health conditions. One report described expected values for daily walking activity in breast cancer patients, and hypothesized that such activity monitoring may facilitate physical activity in this population (De Bruin, E. D., Hartmann, A., Uebelhart, D., Murer, K., & Zijlstra, W. (2008). Although the evidence on the effect of physical activity interventions on daily step activity in cancer survivors has not been established, activity monitors have recently experienced a surge in popularity as a tool for motivating patients and monitoring physical activity. Tudor-Locke, C., McClain, J. J., Hart, T. L., Sisson, S. B., & Washington, T. L. (2009). We conducted a systematic review to summarize the currently available evidence from randomized clinical trials (RCT's) on the effect of physical activity interventions on quantified daily walking behavior in cancer survivors. The secondary aim was to evaluate the methodological quality of these studies.

The word "health'' generally means the state of being free form disease .The world health organization defines health as a "state of complete physical, mental and social well being and not merely the absence of disease or infirmity". So health includes physical, mental and social factors .health may therefore be define as the quality of life that renders in individual "Fit to live most and to serve best'' **(Henneth,1976).**

Aristotle the great philosopher observed that "the body is the temple of soul and to reach harmony of body, mind and spirit, the body must be physically fit '' **(John, 1968).**

Greenberg and pargman opine that health consists of five components – social, mental, emotional, spiritual and physical, when these components are at its balance, a high level of "wellness'' is achieved .physical fitness programmes should be so designed as to achieve this balance and improve each components of health . **(Jerrold, 1986).**

Lzack walron says Look at your health; and if you have it Praise god, and value it next to a good conscience; for health is the second blessing that we mortals are capable of a blessing that money cannot buy **(Lzack walron, 1976).**

Many women find it hard to find the time to exercise due to the demands of parenting and lack of time, money or motivation. However, there are lots of ways you can incorporate exercise into your weekly schedule. A few minor changes to your daily lifestyle can also increase your physical activity level. Regular physical activity is vital for good physical and mental health. It helps improve your overall health and fitness, maintain a healthy weight, reduce your risk for many chronic diseases and promote good mental health. The National Physical Activity Guidelines for Australians recommend that at least 30 minutes of moderate-intensity physical activity on most, preferably all, days is required for good health. This is the same for women and men. However, only 54 per cent of Australian women meet these guidelines. Some of the barriers women face includes caring responsibilities, body image and perceptions of safety. **(Physical activity, 2014)**

Physical activity or exercise can improve your health and reduce the risk of developing certain conditions or diseases including depression and anxiety. Physical activity can help you manage your weight and prevent osteoporosis. Physical activity or exercise can improve your health and reduce the risk of developing several diseases like type 2 diabetes, cancer and cardiovascular disease. Physical activity and exercise can have immediate and long-term health benefits. Most importantly, regular activity can improve your quality of life. A minimum of 30 minutes a day can allow you enjoy these benefits. **(Physical activity, 2014)**

Many factors influence our individual health and that of our communities. Some factors such as socioeconomic status or employment conditions are difficult to change. On the other hand, we have greater control over certain health-related behaviours, such as our eating habits and diet, which when improved can have a positive impact on our overall health. **(Physical activity, 2014)**

Regular exercise and physical activity are important to the physical and mental health of almost everyone, including older adults. Being physically active can help you continue to do the things you enjoy and stay independent as you age. Regular physical activity over long periods of time can produce long-term health benefits. That's why health experts say that older adults should be active every day to maintain their health. In addition, regular exercise and physical activity can reduce the risk of developing some diseases and disabilities that develop as people grow older. In some cases, exercise is an effective treatment for many chronic conditions. For example, studies show that people with arthritis, heart disease, or diabetes benefit from regular exercise. Exercise also helps people with high blood pressure, balance problems, or difficulty walking. One of the great things about physical activity is that there are so many ways to be active. For example, you can be active in short spurts throughout the day, or you can set aside specific times of the day on specific days of the week to exercise. Many physical activities—such as brisk walking, raking leaves, or taking the stairs whenever you can—are free or low cost and do not require special equipment. You could also check out an exercise video from the library or use the fitness center at a local senior center. This guide shows you many types of exercise and physical activity. It also has lots of tips to help you be active in ways that suit your lifestyle, interests, health, and budget, whether you're just starting out, getting back to exercising after a break, or fit enough to run a 3-mile race. It's for everyone—people who are healthy and those who live with an ongoing health problem or disability. **(Physical activity, 2014)**

The aim of this health status report is to provide a "snap-shot" of the health of the population living within the Eastern Health region. To realize Eastern Health's vision, *Healthy People, Healthy Communities*, it is important to have high quality, relevant information that can be used in decision making and evaluation. This website is a tool to improve awareness of the health status of our population, to fill-in existing knowledge gaps, and to help identify pathways for interpreting and sharing population health information (Health Status, 2014).

During the transition from childhood to adulthood, adolescents establish patterns of behavior and make lifestyle choices that affect both their current and future health. Serious health and safety issues such as motor vehicle crashes, violence, substance use, and risky sexual behaviors can adversely affect adolescent and young adults. Some adolescents also struggle to adopt behaviors that could decrease their risk of developing chronic diseases in adulthood, such as eating nutritiously, engaging in physical activity, and choosing not to use tobacco. Environmental factors such as family, peer group, school, and community characteristics also contribute to adolescents' health and risk behaviors. Young people's behaviors are influenced at the individual, peer, family, school, community, and societal levels. Because many sectors of society contribute to adolescent health, safety, and well-being, a collaborative effort that engages multiple partners is necessary. Such joint efforts can also help to promote a more comprehensive approach to addressing adolescent health—one that views each adolescent as a whole person, recognizing and drawing upon his or her assets and not just focusing on risks. To have the most positive impact on adolescent health, government agencies, community organizations, schools, and other community members must work together in a comprehensive approach. Providing safe and nurturing environments for our nation's youth can help ensure that adolescents will be healthy and productive members of society. **(Health Status, 2014)**

The first edition of the web-based Benton County Health Status Report was made available in April 2010. Shortly after the report was made available, Benton County Public Health assessed readability and usability of the report. Benton County conducted a focus group, key informant interviews and an online survey with community members, as well as consulted with health literacy experts, to gather feedback on readability, data utilization, content, web-design, and data presentation. The website is currently undergoing structural changes to address the issues that arose from the usability assessment and the report is being updated with new data following health literacy guidelines. The web-based Health Status Report was developed in response to the need for current county level health data. The purpose of this report is to describe the county's health. It is also intended to facilitate community involvement and cross discipline coordination in addressing public health issues through data-driven decisions and evidence-based practices. This report is a compilation of secondary and primary health indicator data for Benton County. It compares Benton county health indicator data, Oregon health indicator data, and Healthy People targets. The report includes charts and tables, a narrative section that introduces the topic, interprets the data, and provides a prevention message where applicable, and data sources. The Health Status Report is intended to be an ongoing work in progress, not a single cross-sectional glimpse. Information will be updated on a continuing basis as data becomes available. **(Health Status, 2014)**

In adults aged 18–64, physical activity includes leisure time physical activity (for example: walking, dancing, gardening, hiking, swimming), transportation (e.g. walking or cycling), occupational (i.e. work), household chores, play, games, sports or planned exercise, in the context of daily, family, and community activities. In order to improve cardiorespiratory and muscular fitness, bone health, reduce the risk of NCDs and depression: Adults aged 18–64 should do at least 150 minutes of moderate-intensity aerobic physical activity throughout the week or do at least 75 minutes of vigorous-intensity aerobic physical activity throughout the week or an equivalent combination of moderate- and vigorous-intensity activity. Aerobic activity should be performed in bouts of at least 10 minutes duration. For additional health benefits, adults should increase their moderate-intensity aerobic physical activity to 300 minutes per week, or engage in 150 minutes of vigorous-intensity aerobic physical activity per week, or an equivalent combination of moderate- and vigorous-intensity activity. Muscle-strengthening activities should be done involving major muscle groups on 2 or more days a week**. (Physical activity level, May 2014)**

These recommendations are relevant to all healthy adults aged 18–64 years unless specific medical conditions indicate to the contrary. They are applicable for all adults irrespective of gender, race, ethnicity or income level. They also apply to individuals in this age range with chronic nonommunicable conditions not related to mobility such as hypertension or diabetes. These recommendations can be valid for adults with disabilities. However adjustments for each individual based on their exercise capacity and specific health risks or limitations may be needed. There are multiple ways of accumulating the total of 150 minutes per week. The concept of accumulation refers to meeting the goal of 150 minutes per week by performing activities in multiple shorter bouts, of at least 10 minutes each, spread throughout the week then adding together the time spent during each of these bouts: e.g. 30 minutes of moderate-intensity activity 5 times per week. Pregnant, postpartum women and persons with cardiac events may need to take extra precautions and seek medical advice before striving to achieve the recommended levels of physical activity for this age group. Inactive adults or adults with disease limitations will have added health benefits if moving from the category of “no activity” to “some levels” of activity. Adults who currently do not meet the recommendations for physical activity should aim to increase duration, frequency and finally intensity as a target to achieving them. **(Physical activity level, May 2014)**

Physical activity is defined as any bodily movement produced by skeletal muscles that require energy expenditure. Physical inactivity has been identified as the fourth leading risk factor for global mortality causing an estimated 3.2 million deaths globally. Regular moderate intensity physical activity – such as walking, cycling, or participating in sports – has significant benefits for health. For instance, it can reduce the risk of cardiovascular diseases, diabetes, colon and breast cancer, and depression. Moreover adequate levels of physical activity will decrease the risk of a hip or vertebral fracture and help control weight. **(Physical activity level, May 2014)**

Physical activity is any body movement that works your muscles and requires more energy than resting. Walking, running, dancing, swimming, yoga, and gardening are a few examples of physical activity. According to the Department of Health and Human Services' "2008 Physical Activity Guidelines for Americans," physical activity generally refers to movement that enhances health. Exercise is a type of physical activity that's planned and structured. Lifting weights, taking an aerobics class, and playing on a sports team are examples of exercise. Physical activity is good for many parts of your body. This article focuses on the benefits of physical activity for your heart and lungs. The article also provides tips for getting started and staying active and it discusses physical activity as part of a heart healthy lifestyle. **(Physical activity level, May 2014)**

The scientific evidence available for the age group 5–17 years supports the overall conclusion that physical activity provides fundamental health benefits for children and youth. Appropriate levels of physical activity contribute to the development of:

- Healthy musculoskeletal tissues (i.e. bones, muscles and joints);
- Healthy cardiovascular system (i.e. heart and lungs);
- Neuromuscular awareness (i.e. coordination and movement control); and
- It also facilitates maintenance of a healthy body weight.

Moreover, physical activity has been associated with psychological benefits in young people by: improving their control over symptomps of anxiety and depression; and assisting in social development by providing opportunities for self-expression, building self-confidence, social interaction and integration. These recommendations are relevant to all healthy children aged 5–17 years, unless specific medical conditions indicate to the contrary, irrespective of gender, race, ethnicity, or income level. Whenever possible, children and youth with disabilities should meet these recommendations. However they should work with their health care provider to understand the types and amounts of physical activity appropriate for them considering their disability. If children are currently doing no physical activity, doing amounts below the recommended levels will bring more benefits than doing none at all. They should start with small amounts of physical activity and gradually increase duration, frequency and intensity over time.The concept of accumulation refers to meeting the goal of 60 minutes per day by performing activities in multiple shorter bouts spread throughout the day (e.g. 2 bouts of 30 minutes), then adding together the time spent during each of these bouts. **(Physical activity, 2011)**

For children and young people, physical activity includes play, games, sports, transportation, chores, recreation, physical education, orplanned exercise, in the context of family, school, and community activities. The recommendations to improve cardiorespiratory and muscular fitness, bone health, and cardiovascular and metabolic health biomarkers are: Children and youth aged 5-17 should accumulate at least 60 minutes of moderate - to vigorous-intensity physical activity daily. Amounts of physical activity greater than 60 minutes provide additional health benefits. Most of the daily physical activity should be aerobic. Vigorous-intensity activities should be incorporated, including those that strengthen muscle and bone, at least 3 times per week.

The guidelines recommend the types and amounts of physical activity that children, adults, older adults, and other groups should do. The guidelines also provide tips for how to fit physical activity into your daily life.

Children and youth do 60 minutes or more of physical activity every day. Activities should vary and be a good fit for their age and physical development. Children are naturally active, especially when they're involved in unstructured play (like recess). Any type of activity counts toward the advised 60 minutes or more. Most physical activity should be moderate-intensity aerobic activity. Examples include walking, running, skipping, playing on the playground, playing basketball, and biking. Vigorous-intensity aerobic activity should be included at least 3 days a week. Examples include running, doing jumping jacks, and fast swimming. Muscle-strengthening activities should be included at least 3 days a week. Examples include playing on playground equipment, playing tug-of-war, and doing pushups and pull ups. Bone-strengthening activities should be included at least 3 days a week. **(Physical activity, 2014)**

Examples include hopping, skipping, doing jumping jacks, playing volleyball, and working with resistance bands. Children and youth who have disabilities should work with their doctors to find out what types and amounts of physical activity are safe for them. When possible, these children should meet the recommendations in the guidelines. Some experts also advise that children and youth reduce screen time because it limits time for physical activity. They recommend that children aged 2 and older should spend no more than 2 hours a day watching television or using a computer (except for school work). **(Physical activity, 2014)**

1.3 Guidelines for Adults

Some physical activity is better than none. Inactive adults should gradually increase their level of activity. People gain health benefits from as little as 60 minutes of moderate-intensity aerobic activity per week. For major health benefits, do at least 150 minutes (2 hours and 30 minutes) of moderate-intensity aerobic activity or 75 minutes (1 hour and 15 minutes) of vigorous-intensity aerobic activity each week. Another option is to do a combination of both. A general rule is that 2 minutes of moderate-intensity activity counts the same as 1 minute of vigorous-intensity activity. For even more health benefits, do 300 minutes (5 hours) of moderate-intensity aerobic activity or 150 minutes (2 hours and 30 minutes) of vigorous-intensity activity each week (or a combination of both). The more active you are, the more you will benefit. When doing aerobic activity, do it for at least 10 minutes at a time. Spread the activity throughout the week. Muscle-strengthening activities that are moderate or vigorous intensity should be included 2 or more days a week. These activities should work all of the major muscle groups (legs, hips, back, chest, abdomen, shoulders, and arms).

Examples include lifting weights, working with resistance bands, and doing sit-ups and push-ups, yoga, and heavy gardening. **(Physical activity, 2014)**

1.4 Guidelines for Adults Aged 65 or Older

Older adults should be physically active. Older adults who do any amount of physical activity gain some health benefits. If inactive, older adults should gradually increase their activity levels and avoid vigorous activity at first. Older adults should follow the guidelines for adults, if possible. Do a variety of activities, including walking. Walking has been shown to provide health benefits and a low risk of injury. If you can't do 150 minutes (2 hours and 30 minutes) of activity each week, be as physically active as your abilities and condition allow. You should do balance exercises if you're at risk for falls. Examples include walking backward or sideways, standing on one leg, and standing from a sitting position several times in a row. If you have a chronic (ongoing) condition—such as heart disease, lung disease, or diabetes—ask your doctor what types and amounts of activity are safe for you. **(Physical activity, 2014)**

1.5 Guidelines for Women During Pregnancy and Soon After Delivery

You should ask your doctor what physical activities are safe to do during pregnancy and after delivery. If you're healthy but not already active, do at least 150 minutes (2 hours and 30 minutes) of moderate-intensity aerobic activity each week. If possible, spread this activity across the week. If you're already active, you can continue being active as long as you stay healthy and talk with your doctor about your activity level throughout your pregnancy. After the first 3 months of pregnancy, you shouldn't do exercises that involve lying on your back. You shouldn't do activities in which you might fall or hurt yourself, such as horseback riding, downhill skiing, soccer, and basketball. **(Physical activity, 2014)**

Twenty minutes of daily physical activity are critical to making publicly funded schools healthier places to learn and improve student achievement. This step in the government's Healthy Schools Plan will ensure that elementary students have a minimum of 20 minutes of sustained moderate to vigorous physical activity each school day. On October 6, 2005, Education Minister Gerard Kennedy announced that every elementary student will take part in a minimum of 20 minutes of daily physical activity as part of the government's Healthy Schools Program. **(Daily physical activity, 2014)**

This is the first Surgeon General's report to address physical activity and health. The main message of this report is that Americans can substantially improve their health and quality of life by including moderate amounts of physical activity in their daily lives. Health benefits from physical activity are thus achievable for most Americans, including those who may dislike vigorous exercise and those who may have been previously discouraged by the difficulty of adhering to a program of vigorous exercise. For those who are already achieving regular moderate amounts of activity, additional benefits can be gained by further increases in activity level. **(Physical activity and health, April 2014)**

This report grew out of an emerging consensus among epidemiologists, experts in exercise science, and health professionals that physical activity need not be of vigorous intensity for it to improve health. Moreover, health benefits appear to be proportional to amount of activity; thus, every increase in activity adds some benefit.

Emphasizing the amount rather than the intensity of physical activity offers more options for people to select from in incorporating physical activity into their daily lives. Thus, a moderate amount of activity can be obtained in a 30- minute brisk walk, 30 minutes of lawn mowing or raking leaves, a 15-minute run, or 45 minutes of playing volleyball, and these activities can be varied from day to day. It is hoped that this different emphasis on moderate amounts of activity, and the flexibility to vary activities according to personal preference and life circumstances, will encourage more people to make physical activity a regular and sustainable part of their lives. The information in this report summarizes a diverse literature from the fields of epidemiology, exercise physiology, medicine, and the behavioral sciences. The report highlights what is known about physical activity and health, as well as what is being learned about promoting physical activity among adults and young people. **(Physical activity and health April 2014)**

Many women find it hard to find the time to exercise due to the demands of parenting and lack of time, money or motivation. However, there are lots of ways you can incorporate exercise into your weekly schedule. A few minor changes to your daily lifestyle can also increase your physical activity level. Regular physical activity is vital for good physical and mental health. It helps improve your overall health and fitness, maintain a healthy weight, reduce your risk for many chronic diseases and promote good mental health.

The National Physical Activity Guidelines for Australians recommend that at least 30 minutes of moderate-intensity physical activity on most, preferably all, days is required for good health. This is the same for women and men. However, only 54 per cent of Australian women meet these guidelines. Some of the barriers women face includes caring possibilities, body image and perceptions of safety. **(Better health, April 2014)**

Culturally, the abbreviation "PMS" is widely understood in English-speaking countries to refer to difficulties associated with menses, and the abbreviation is used frequently even in casual and colloquial settings, without regard to medical rigor. In these contexts, the syndrome is rarely referred to without abbreviation, and the connotations of the reference are frequently more broad than the clinical definition. Premenstrual dysphoric disorder (PMDD) consists of symptoms similar to, but more severe than, PMS, and while primarily mood-related, may include physical symptoms such as bloating. PMDD is classified as a repeating transitory cyclic disorder with similarities to unipolar depression, and several antidepressants have been approved as therapy. **(Premenstrual Syndrome, May 2014)**

PMS most commonly affects women aged 30-40 years. However, it can affect you at any age, even if you are less than 20 years old. Most women can tell that a period is due by the way they feel both physically and mentally. For most, the symptoms are mild and not troublesome. About one woman in twenty has PMS where the symptoms become bad enough to disrupt normal functioning and quality of life. Day-to-day life and performance at work can be affected. It may cause tension with family and friends. **(Premenstrual Syndrome, May 2014)**

Premenstrual syndrome (PMS) is the name given to the physical, psychological and behavioural symptoms that can occur in the two weeks A before a woman's monthly period. It is also known as premenstrual tension (PMT). There are many different symptoms of PMS, but typical examples are fluid retention, breast tenderness, mood swings, feeling irritable. Usually, these symptoms improve when the woman's period starts and disappear a few days afterwards. A small number of women find that their symptoms are severe enough to stop them living their normal lives. This is due to a more intense type of PMS known as premenstrual dysphoric disorder. **(Premenstrual Syndrome, May 2014)**

Nearly all women of child-bearing age have some premenstrual symptoms, butÂ those between their late 20s and early 40s are most likely to experience PMS. The exact cause of PMS is not fully understood. However, it is thought to be linked to the changing levels of hormones in the body during a woman's menstrual cycle. There is no cure for PMS, but there are treatments available and changes that can be made to help women manage their symptoms. PMS usually improves after the menopause. **(Premenstrual Syndrome, May 2014)**

Premenstrual syndrome, or PMS, refers to the range of physical and emotional symptoms that many women experience in the lead up to a period (menstruation). Symptoms resolve once the woman's period begins and there is usually at least one symptom-free week before the symptoms return. It is thought that most menstruating women have premenstrual symptoms, ranging from relatively mild (in 75 per cent of women) to severe (in 20 to 30 per cent of women). In 8 per cent of women with severe symptoms, PMS is associated with reduced quality of life.PMS is a complex condition that includes physical and emotional symptoms. The latest research points to changes in brain chemicals (neurotransmitters) in the time after ovulation and before menstruation. Life stressors and a genetic link may also play a role. Although the cause isn't conclusively known, PMS can be managed with various medications and other strategies. **(Better health, 2014)**

PMS is common in young and middle-aged women, and three in four women have some PMS symptoms. It usually starts in the second half of your menstrual cycle and goes away by the time your period starts. Older teenage girls tend to have more severe symptoms than younger teenage girls, and women in their 40s tend to be affected most severely.PMS can range in severity from mild to severe. Mild PMS causes some discomfort, but doesn't interfere with your life. If you have moderate PMS, your symptoms interfere with your personal, social and professional life, but you're still able to function each day, although maybe not to your usual level. If you have severe PMS, your symptoms are so severe that the condition affects your quality of life and relationships. A more severe form of PMS is premenstrual dysphoric disorder. Emotional symptoms, such as mood swings, depression and irritability, are more common in premenstrual dysphoric disorder and the condition can interfere significantly with your everyday life. It affects between three and eight women in every 100. **(Premenstrual Syndrome, 2014)**

If you keep a symptom diary, you may find that your physical or emotional symptoms are linked with your period. You can then predict more accurately how you will feel at certain times of the month. This may help you to plan your time so you can try, wherever possible, to prevent being in stressful situations on key days and pinpoint any emotional triggers that make your symptoms worse. Regular exercise may help to improve your symptoms. Aim to do at least 150 minutes (two and a half hours) of moderate intensity physical activity during the week. Your GP may advise you to eat a healthy, balanced diet that's low in saturated fat, sugar and salt, and high in fibre, vegetables and fruit. You may also find that eating smaller meals more regularly throughout the day when you have PMS will help to reduce your symptoms. Some research suggests that reducing the amount of salt you eat may minimise the bloated feeling and tender breasts that you may have with PMS. Other studies recommend limiting caffeine because it's related to premenstrual irritability and insomnia. However, more research is needed to prove these effects. There is no scientific evidence to say that taking vitamin B6 supplements is helpful for PMS, and high doses can damage your nervous system. **(Premenstrual Syndrome, 2014)**

Women's menstrual cycles are controlled by a complex interaction of female hormones. These hormones help initiate menstruation during puberty, determine the rhythm and length of menstrual periods during childbearing years, and signal the end of menstruation at menopause. Hormonal control of menstruation involves the brain, pituitary gland, and ovaries. The exact cause of PMS is unknown. It's thought to be related to changes in the level of specific hormones. A few studies suggest that PMS symptoms are linked to premenstrual fluctuations in a brain chemical called *serotonin* and increased sensitivity to the hormone progesterone. Other studies suggest that the hormone estrogen causes fluid retention, which probably explains the temporary weight gain, breast tenderness, and bloating experienced by many women with PMS. Recent research suggests that women with PMS may metabolize progesterone differently. Other hormonal and metabolic changes may also be involved, but further research is needed. **(Premenstrual Syndrome, 2014)**

Proper nutrition, regular exercise, and a healthy lifestyle generally help with PMS. Speaking with your doctor may help remove fears surrounding this diagnosis. A combination of one or more of the following treatments may help relieve symptoms:

Medications: Taking oral contraceptives may help stabilize the changes in hormone levels and stop ovulation. Bloating and water retention can be improved by cutting down on salt and by using a mild diuretic that will make you urinate. Taking over-the-counter pain relievers such as acetaminophen or ibuprofen can help relieve headaches, joint pain, and menstrual cramping. The antidepressant medications known as selective serotonin reuptake inhibitors (SSRIs) are widely used to manage psychological symptoms like anxiety and depression. These antidepressants have been proven to help manage PMS. However, you need to discuss this with your doctor first before starting any medication or herbal remedy to make sure it's appropriate and safe for you. **(Premenstrual Syndrome, 2014)**

Changes in diet: Cutting back on sugar, coffee, and alcohol may provide some relief from symptoms. Taking supplements containing calcium, magnesium, soy, vitamin B6, and vitamin E may reduce certain symptoms. To date, among these supplements, calcium has the most convincing evidence. Some women use certain herbal products such as evening primrose oil, St. John's wort, and chasteberry to decrease their symptoms; however, the effectiveness of such products is not known. Always talk to your doctor or pharmacist before taking a vitamin, supplement, or herbal product. Regular exercise and stress reduction techniques may decrease nervousness and agitation as well as reduce symptoms associated with PMS. Speaking to a counsellor may help some women cope better with the psychological effects of PMS. Keeping a symptom diary can give you and your doctor a better picture of your symptoms and help you evaluate the effects of different treatments. **(Premenstrual Syndrome, 2014)**

Physical activity is any body movement produced by muscles and that results in the release of energy. Physical activity helps build and maintain strong muscles, bones, and joints. It reduces the risk for heart disease, some cancers, diabetes, and helps prevent unwanted weight gain. Physical activity also reduces stress and anxiety, increases a sense of well-being and improves our quality of life. People of all ages can improve their health by being moderately active on most days of the week. Examples of moderate activity include brisk walking, gardening, and yard work sufficient to increase the heart rate. **(Physical activity intervention, 2014)**

Increasing awareness and knowledge of the importance of physical activity is an important first step for becoming active. Developing skills and improving motivation to be physical activity on a regular basis are important as well. Creating supportive and safe environments in which physical activity can take place are essential for increasing physical activity in a community. Included in the Intervention MICA physical activity topic are considerations for implementing interventions with different populations and in different settings. Evidence-based strategies and interventions are described in detail. Action steps, tools and resources for planning and implementing a physical activity intervention are provided. **(Physical activity intervention, 2014)**

1.6 STATEMENT OF THE PROBLEM

The overall purpose of this study was to analysis the level of physical activity, premenstrual syndrome, health status and physical activity intervention among the college women students

1.7 RESEARCH QUESTIONS

The purpose of this study was to analysis the level of physical activity, premenstrual syndrome, health status and physical activity intervention among the college women students.

- To analyse the level of physical activity among college women students
- To compare the premenstrual syndrome among college women students
- To analyse the health status among college women students
- To determine the influence exercise intervention programme on health status and premenstrual syndrome among college women students.

1.8 HYPOTHESES

It was hypothesised that;

1. There would be significant difference on level of physical activity, premenstrual syndrome and health status among college level women students.
2. There would be significant difference on level of physical activity, premenstrual syndrome and health status among college level women students.
3. There would be significant improvement on premenstrual syndrome and health status due to the effects of physical activity intervention.
4. There would be significant difference between experimental and control groups between experimental and control groups.

1.9 ASSUMPTIONS

Since the study is conducted based on survey research, it is presumed that respondents clearly understood the questionnaire and thoroughly and honestly answered each survey question.

1.10 DELIMITATIONS

1. A pilot study was conducted to determine the effectiveness and delivery of the questionnaire to be administered
2. For the present analysis only the adolescences women were selected.

3. The selected variables were tested by standardized questionnaire.
4. To achieve the purpose of the study, initially 1000 female were selected from the affiliated colleges of Manonmaniam Sundaranar University, Tirunelveli at random to analyse the level of physical activity, premenstrual syndrome and health status.
5. From 1000 participants, 200 participants from each college were selected at random.
6. For the exercise intervention programme, 30 students were selected from Rani Anna Government College for Women, Tirunelveli were selected as subjects at random.
7. Stratified random sampling method was adopted for this study. Five women's colleges were selected from Tirunelveli districts such as Rani Anna College–Tirunelveli, Sarah Tucker College–Tirunelveli, Sri Saratha College–Tirunelveli, Rose Mary College-Tirunelveli and Sri Parasakthi College-Tirunelveli. Two hundred students were selected at random from each college.
8. Level of physical activity, premenstrual syndrome and health status were selected as criterion variables and exercise intervention programme was selected as independent variables.
9. The age of the subjects ranged from and 18 to22 years.

1.11 LIMITATIONS

The following limitations were not considered while interpreting the results of the study, they are

1) The previous experience of the subject in the field of sports and games, which might be influencing on the data collection, was not considered.
2) Psychological factors, food habits rest period, life style and so forth could not be controlled.
3) The weather condition such as atmospheric temperature, humidity and Meteorological factors during testing period were also not considered.
4) Through the subjects were motivated verbally, no attempt was made to differentiate the motivation level during the period of testing.
5) The data were collected using a survey.
6) The participants to response honestly and in the timely manner to the survey directly influenced the result.
7) The availability of participants directly influenced the result.
8) The questionnaire method was a adopted for this study.

1.12 SIGNIFICANCE OF THE STUDY

1) This study would help to find out the availability of health tools.
2) This study would help to find out the level of physical activity in college women student.

3) The results of the present study will also be helpful for future studies related to this area.

4) This study will be helpful to know about the use of physical activity among college women students.

5) The finding of the study will add to the existing knowledge with regard to the selected variables among college students.

6) The finding of the study will add to the quantum of knowledge in the area of health status among college students.

7) The study may help to assess the premenstrual syndrome among college students.

8) A unique aspect of this work is that it includes recommendations for the practical use of the research findings.

1.13 DEFINITIONS OF THE OPERATIONAL TERMS

Physical activity

Physical activity can be defined as any movement of the body that requires energy expenditure. This includes any motion you do through the day excluding sitting still or lying down. For example, walking to class, taking the stairs, mowing the lawn, and even cleaning your house can be considered physical activity. Exercise is a type of physical activity but not every physical activity is exercise. Exercise is a planned, structured, and repetitive activity for the purpose of improving or maintain physical fitness.

Premenstrual syndrome

A combination of physical and mood disturbances that occur in the last half of a woman's menstrual cycle after ovulation which normally end with the onset of the menstrual flow. Physical features of the premenstrual syndrome (PMS) include breast tenderness and bloating. Psychological changes include anger and depression.

Health status

Health status is the current state of your own health, It includes the status of your wellness, fitness, and any underlying diseases or injuries, **(Health status, 2010).**

1.14 ORGANIZATION OF THE DISSERTATION

This dissertation includes five chapters. The first chapter is an introductory chapter that provides a brief explanation of the sports training, purposes of the study, boundaries of the study, generation of research hypotheses and significance of the study. The second chapter is the literature review. It introduces information about Internet usage patterns, and describes the theoretical under pinning's for the proposed model. The third chapter describes research methodology. It consists of training and testing procedures, measures, sample selection, and data analysis procedures.

In the fourth chapter, analysis of the collected data was presented. It includes results of measurement and structural model tests, and hypotheses tests.

The fifth chapter discusses the findings and conclusions of the research, as well as implications and recommendations for future research.

Chapter II

REVIEW OF RELATED LITERATURE

2.1 Introduction

A study of relevant literature is an essential step to get a full picture of what has been done with regard to the problem under the study. Such a review brings new ideas, theories, comparative materials and helps in the development of research procedure.

The research scholar had come across general books, periodicals, journals, internet and unpublished theses while searching for relevant facts and findings that are related to his present study. Such of these facts are given below for a better understanding and to justify her study.

The phrase “review of literature” consists of two words “review” and “literature” .In research methodology the term literature refers to the knowledge of a particular area of any discipline which includes theoretical practical and its research studies.

The literature is any field forms the foundation upon which the future work will be built. If we fail to build foundation of knowledge provided by the review of literature one work is likely to be shadow and that has already done better by someone else.

2.2 Studies on Physical activity barrier, Premenstrual syndrome and Health status

Ahangar, K. S., Dolatshahei, B., & Badeie, M. M. (2011) studied is a kind of practical researches with anterior and posterior test to control group. first, 150 individuals of women crowd, has completed a daily menstrual symptom diary for two consecutive months; then within them, final sample ($n = 30$) based on women that had been possessed much more anger symptoms samples divided in two accidential groups: receiving group with 12 lessons 2 hours per each one for anger management training ($n = 15$) and non-receiving group for training ($n = 15$). Spielberger, anger test 2 (STAXI-2) have been utilized for all of participants (test and control) in posterior and anterior form .This propaganda allow us to evaluate profiting of anger management training to reduce anger severity symptoms of PMS.There in a meaningful discrepancy in anger severity symptoms of PMS before and after anger management training.12 weekly training lessons for anger management decreased meaningful anger severity symptoms of PMS.

Al-Isa, A. N., Campbell, J., Desapriya, E., & Wijesinghe, N. (2011) explored the social and health factors that are associated with the level of physical activity among Kuwaiti college students. A random sample of 787 students (48% males and 52% females) was chosen and weight and height were measured to obtain body mass index (BMI, kg/m(2)). Associated social and health factors were obtained using a questionnaire. Those reporting being physically inactive numbered 354 and the remaining 433 were active. Obesity among males was 13% and was 10.5% among females. The social and health factors that were found to be significantly associated with physical activity among the students were gender ($P < .001$), marital status ($P < .05$), BMI category (obese or nonobese) ($P < .05$), last dental and health checkup ($P < .01$), desiring a higher degree ($P < .001$), and countries preferred for visiting ($P < .01$). Males significantly exceeded females in the practice of physical activity. In conclusion, behavioural modifications, intervention studies, and health education touting the benefits of being physically active should be instituted to increase the practice of sports and other physical activities in order to control and decrease obesity-related morbidity and mortality.

Al-Kandari, F., & Vidal, V. L. (2007) studied of 224 nursing students assessed their health-promoting lifestyle profile and correlated it with the levels of enrollment in nursing courses and academic performance. The health-promoting lifestyle profile was measured by Walker's Health-promoting Lifestyle Profile II instrument. Academic performance was measured by assessing the nursing grade point average and general grade point average of the students. The students had positive health-promoting lifestyles with significant differences noted between males and females in the overall profile, physical activity, interpersonal relations, and stress management. Sociodemographic variables, such as age, nationality, and marital status, but not income, showed an association with students' health-promoting lifestyles. A significant correlation was noted between students' nursing enrollment and level of health responsibility. No significant correlation was established between a health-promoting lifestyle and academic performance. This study poses a challenge for nurse educators to provide an effective environment to maximize students' potential to be future vanguards of health.

Allais, G., Gabellari, I. C., Burzio, C., Rolando, S., De Lorenzo, C., Mana, O., & Benedetto, C. (2012) determined Premenstrual syndrome (PMS) includes a wide variety of physical, psychological, and cognitive symptoms that occur recurrently and cyclically during the luteal phase of the menstrual cycle and disappear soon after the onset of menstruation. Headache, often of migrainous type, is one of physical symptoms often reported in the diagnostic criteria for PMS. Menstrual migraine (MM) is a particular subtype of migraine occurring within the 2 days before and the 3 days after the onset of menses. According to this definition, therefore, some attacks of MM certainly occur in conjunction with the period of maximum exacerbation of PMS symptoms. The relationship between MM and PMS has been investigated through diary-based studies which have confirmed the possible correlation between these two conditions. In this paper we provide indications for the treatment of MM, making particular reference to those therapies that may be useful in the treatment of PMS symptoms. Even if triptans are the gold standard for the acute treatment, if symptomatic treatment is not sufficient one can resort to a short-term perimenstrual prophylaxis. Non-steroidal anti-inflammatory drugs have been demonstrated effective in MM prophylaxis. Among natural products there is some evidence of efficacy for magnesium, phytoestrogens, and ginkgolide B. Finally, also a combined oral contraceptive containing drospirenone, taken continuously for 168 days, has shown promising results.

Allen, L.M., Lam, A.C. (2012) examined PMS/PMDD is a pervasive problem with a significant impact on the quality of life of affected individuals. This condition most often begins in adolescence with the establishment of normal ovulatory menstrual cycles; however, the underlying pathophysiology has yet to be delineated. Prospective evaluation is key to the confirmation of the diagnosis before the initiation of pharmacotherapy, especially psychotropic therapies, due to the possibly harmful side effect profile for adolescents compared to adults. Similarly, dysmenorrhea is common in adolescents. Although the majority of cases are primary, the HCP must be vigilant to allow for early diagnosis and treatment of secondary causes, thereby preventing long-term sequelae of delayed diagnosis. Stepwise therapy for dysmenorrhea treatment is usually employed; the choice of therapy should account for contraceptive needs of the adolescent in addition to symptom relief. For both PMSS/PMDD and dysmenorrhea, most studies of therapy have been performed in adults and, as such, adolescent-specific trials are required to confirm applicability to this age group.

Allender, S., Cowburn, G., & Foster, C. (2006) aimed to explore the participants' experiences of sport and physical activity and reasons for participation or non-participation in sport and physical activity, collected information on participants who lived in the United Kingdom and presented data collected using qualitative methods. From >1200 papers identified in the initial search, 24 papers met all inclusion criteria. The majority of these reported research with young people based in community settings. Weight management, social interaction and enjoyment were common reasons for participation in sport and physical activity. Concerns about maintaining a slim body shape motivated participation among young girls. Older people identified the importance of sport and physical activity in staving off the effects of aging and providing a social support network. Challenges to identity such as having to show others an unfit body, lacking confidence and competence in core skills or appearing overly masculine were barriers to participation.

Andegiorgish, A. K., Wang, J., Zhang, X., Liu, X., & Zhu, H. (2012) estimated the prevalence and determinants of overweight and obesity among school children and adolescents (7-18 years), a cross-sectional study was conducted in Tianjin City of Northeast China. Five primary and middle schools were selected using a multistage random cluster sampling. Anthropometric measurements were taken by the research team. Then, standardized questionnaires on socioeconomic status, diet and lifestyle, parental weight and height were sent to the parents to be completed together with their child. Overall, 3,140 students were screened for overweight and associated risk factors. Body mass index classification reference proposed by the Working Group on Obesity in China was used as a screening reference to calculate the prevalence of overweight in these groups. The prevalence of overweight and obesity was found to be 12.5% and 15.7%, respectively. Logistic regression analysis (overweight compared with normal weight) showed that, overweight in children was significantly associated with male gender, parental obesity, parental educational level, mother's history of gestational diabetes, high birth weight, less hours of physical activity per day, urban residence, motorized transportation, and eating food not prepared at home. After adjusting for parental obesity, the odds ratio of overweight children was increased by urban residence (rural as a reference, 2.68, 2.16-3.32), positive gestational diabetes (2.76, 1.37-4.50), and fast-food/restaurant food service (2.03, 1.34-3.07). Children who walked to school and participated in outdoor activities were 54% and 63% less likely to be overweight respectively. The prevalence of overweight and obesity is high among school children and adolescents in Tianjin. Changes in lifestyle, high socioeconomic status, parental obesity, mother's history of gestational diabetes, and high birth weight are among the highly associated risk factors of overweight. An educational approach about maintaining a healthy weight should be introduced much earlier to high-risk children, parents, and women of childbearing age.

Andrade Osorio, É., Padilla Raygoza, N., & Ruiz Paloalto, M. (2013) compared relationship between perceived barriers and physical activity level in older adults from Aguascalientes in Mexico. The present study is cross-sectional, analytic, and observational.150 older adults registered in the Family Integral Development System from Aguascalientes State were included at random, who accepted to participate by signing a consent form. A questionnaire to detect perceived barriers as well as an international physical activity questionnaire was applied. Z for two proportions, p-value and odds ratio, 95% confidence interval were calculated, between external and internal perceived barriers and low physical activity level in older adults from the sample. The nursing designing programs to perform physical activity in older adults should take into account these perceived barriers.

Arbabi, M., Shirmohammadi, M., Taghizadeh, Z., Haghanni, H., & Parsafar, H. (2011) assessed the co-morbidity of PMS andpsychiatric disorders in a sample of girls with PMS compared to those without PMS.This study was conducted through a cross sectional method with 362 participants (166 with PMS and 196 healthy girls) who were selected randomly and completed the demographic questionnaire, premenstrual syndrome symptom daily record scale and the symptom checklist 90-revised (SCL-90-R).There is a considerable relationship between PMS and different psychiatric symptoms that can complicate the diagnosis of PMS and its treatment for the health care providers. Therefore, all health care providers who are in contact with women in their reproductive age should be sensitive to mental health status in women with PMS.

Bale, J. M., Gazmararian, J. A., & Elon, L. (2015) determined whether workload, job satisfaction, and flexible schedules are predictive of using time at work to exercise. The study design was the quantitative analysis of the time at work to exercise intervention of a cluster-randomized controlled trial. The study took place at Emory University, Atlanta, Georgia. Subjects comprised 188 full- and part-time faculty and staff (57% female). Employees were randomized into different intervention groups, and analysis focused on employees who were provided 30 minutes of time at work to exercise. Time at work to exercise at was measured 9 months, and work environment characteristics were measured at baseline. Logistic regression modeling using generalized estimating equation analysis was used to account for departmental clustering. The results support the hypothesis that individuals with a supportive work environment would be more likely to use time at work to exercise. Workload and having time during the day are more important than job satisfaction.

Bastani, F., & Hashemi, S. (2012) conducted to determine the effects of a web based education on general health in students. A total of 104 women with PMS participated in the study and were randomly assigned to two groups. The instrument for data collection was a questionnaire. The web-based training can lead to post-intervention improvements in general health and relief of PMS in female university students.

Basu, S., Seligman, H., & Winkleby, M. (2014) combined a metabolic and an epidemiological model of obesity to estimate changes in calorie intake and physical activity necessary to achieve the Healthy People 2020 objective of reducing adult obesity prevalence from 33.9% to 30.5%.We used the National Health and Nutrition Examination Survey (1999-2010) to construct and validate a microsimulation model of the US population aged 10 years and older, for 2010 to 2020.Changes in calorie intake and physical activity portend rising inequalities in obesity prevalence. These changes require multiple simultaneous population interventions.

Bendich, A. (2001) discussed elucidating the roles of nutrients in optimizing human health included the support of research in the areas of women's health and immune function. Several essential nutrients have been shown to affect women's health throughout the different life stages. Symptoms of premenstrual syndrome affect the vast majority of menstruating women, and calcium supplementation significantly reduces physical and emotional symptoms. Premenstrual syndrome in fact might be a predictor of osteoporosis induced by low calcium intake. Periconceptional multivitamin supplementation has reduced the risk of serious birth defects, premature delivery, and low birth weight by 50% and improved maternal health during pregnancy. Micronutrients of particular importance for prevention of adverse pregnancy outcomes are folic acid, zinc, and iron. However, if the preterm delivery is caused by preeclampsia, then data suggest that calcium supplementation and high doses of vitamins C and E significantly reduce that risk. Well-controlled studies consistently have shown that calcium supplementation, with or without vitamin D, significantly reduces the risk of hip fracture. Antioxidants such as vitamins C and E have been shown to reduce the risk of fracture in women smokers. As in the rapidly growing embryo, the immune system includes rapidly multiplying cells whose functions are dramatically affected by an individual's micronutrient status. Multivitamins have been shown to enhance many aspects of immune response, and antioxidant micronutrients consistently have been found to enhance lymphocyte-proliferative responses and skin-test responses, especially in the elderly.

Berger, G., & Peerson, A. (2009) investigated the relationship between physical activity and the social milieu of young Muslim women in the United Arab Emirates. This participatory action research project included semi-structured in-depth interviews and focus groups and yielded qualitative data. Set within a context of rapid social change, perceived barriers to daily exercise influenced participants' physical activity levels and overall well-being. Results indicated a lack of physical exercise and strategies were proposed for implementation by college staff and students.

Bertone-Johnson, E. R., Chocano-Bedoya, P. O., Zagarins, S. E., Micka, A. E., & Ronnenberg, A. G. (2010) assessed this relationship in a cross-sectional analysis within the UMass Vitamin D Status Study. Between 2006 and 2008, 186 women aged 18–30 (mean age = 21.6 years) completed a validated food frequency questionnaire, additional questionnaires to assess menstrual symptoms and other health and lifestyle factors, and provided a fasting blood sample collected during the late luteal phase of their menstrual cycle. Among all study participants, results suggested the possibility of an inverse association between intake of vitamin D from food sources and overall menstrual symptom severity, though were not statistically significant; mean intakes in women reporting menstrual symptom severity of none/minimal, mild, and moderate/severe were 253, 214, and 194 IU/day, respectively ($P = 0.18$). From among all study participants, 44 women meeting standard criteria for PMS and 46 women meeting control criteria were included in additional case–control analyses. In these women, after adjustment for age, body mass index, smoking status and total calcium intake, higher intake of vitamin D from foods was associated with a significant lower prevalence of PMS. Women reporting vitamin D intake from food sources of $\geq$100 IU/day had a prevalence odds ratio of 0.31 compared to those reporting <100 IU/day (95% confidence interval = 0.10–0.98). Late luteal phase 25-hydroxyvitamin D_3 levels were not associated with prevalent PMS. Results from this pilot study suggest that a relationship between vitamin D and PMS is possible, though larger studies are needed to further evaluate this relationship and to investigate whether 25-hydroxyvitamin D_3 levels in the follicular or early luteal phases of the menstrual cycle may be related to PMS risk.

Bertone-Johnson, E. R., Hankinson, S. E., Bendich, A., Johnson, S. R., Willett, W. C., & Manson, J. E. (2005) conducted a case-control study nested within the prospective Nurses' Health Study II cohort. Participants were a subset of women aged 27 to 44 years and free from PMS at baseline in 1991, including 1057 women who developed PMS over 10 years of follow-up and 1968 women reporting no diagnosis of PMS and no or minimal menstrual symptoms. Intake of calcium and vitamin D was measured in 1991, 1995, and 1999 by a food frequency questionnaire A high intake of calcium and vitamin D may reduce the risk of PMS. Large-scale clinical trials addressing this issue are warranted. Given that calcium and vitamin D may also reduce the risk of osteoporosis and some cancers, clinicians may consider recommending these nutrients even for younger women.

Biggs, W. S., & Demuth, R. H. (2011) defined as recurrent moderate psychological and physical symptoms that occur during the luteal phase of menses and resolve with menstruation. It affects 20 to 32 percent of premenopausal women. Women with premenstrual dysphoric disorder experience affective or somatic symptoms that cause severe dysfunction in social or occupational realms. The disorder affects 3 to 8 percent of premenopausal women. Proposed etiologies include increased sensitivity to normal cycling levels of estrogen and progesterone, increased aldosterone and plasma renin activity, and neurotransmitter abnormalities, particularly serotonin. The Daily Record of Severity of Problems is one tool with which women may self-report the presence and severity of premenstrual symptoms that correlate with the criteria for premenstrual dysphoric disorder in the Diagnostic and Statistical Manual of Mental Disorders, 4th ed., text revision. Symptom relief is the goal for treatment of premenstrual syndrome and premenstrual dysphoric disorder. There is limited evidence to support the use of calcium, vitamin D, and vitamin B6 supplementation, and insufficient evidence to support cognitive behavior therapy. Serotonergic antidepressants (citalopram, escitalopram, fluoxetine, sertraline, venlafaxine) are first-line pharmacologic therapy.

Bodde, A. E., Seo, D. C., & Frey, G. (2009) examined physical activity participation as a potential predictor of self-rated health status of a unique sub-population of non-elderly, non-assistive device using adults with disabilities. This cross-sectional study analyzed population-based data from the 2007 U.S.

Behavioral Risk Factor Surveillance System. A sample of 46,883 adults with disabilities was examined. Multivariable logistic regression analyses were conducted to determine whether physical activity participation predicted self-rated health status controlling for age, gender, income level, health insurance status, BMI, smoking status, education level, and heart disease, arthritis, asthma and diabetes diagnoses. Adequate physical activity participation is correlated to better health status among adults with disabilities not associated with aging. Physical activity promotion is warranted for this population.

Borenstein, J. E., Dean, B. B., Leifke, E., Korner, P., & Yonkers, K. A. (2007) compared healthcare utilization in women with and without PMS according to the Daily Record of Severity of Problems (DRSP) scores in order to determine the degree of symptomatic variation in premenstrual symptoms associated with differences in the burden of illness. Participants were women aged 18-45 years enrolled in a medical group in southern California. Respondents completed the Medical Outcomes Study Short Form-36 (SF-36) at baseline and the DRSP symptom and occupational productivity items daily. Luteal phase DRSP scores were averaged over two consecutive cycles. Respondents were categorized as having mild/no and moderate/severe PMS or premenstrual dysphoric disorder (PMDD) using a previously published, validated algorithm. Medical costs were estimated from medical claims data over the 2 years prior to study entry. A 43% difference in premenstrual vs. postmenstrual symptom scores is associated with a significant difference in healthcare burden. These data support the use of a 50% reduction in symptom ratings as a clinically relevant improvement in PMS/PMDD treatment trials, although smaller differences may also be meaningful.

Boyle, J., Mattern, C. O., Lassiter, J. W., & Ritzler, J. A. (2011) determine if a course-based, peer education intervention was associated with increases in PA and physical fitness. Participants were 178 students enrolled in a personal health class during the 2007-2008 academic year. A quasi-experimental design was used to assess the effect of the intervention. Repeated measures analysis of covariance was employed to test the impact of the intervention on students' PA, body composition, waist-to-hip ratio, cardiovascular fitness, flexibility, and muscular strength. The intervention was effective in improving PA and physical fitness among college women.

Boyle, S. E., Jones, G. L., & Walters, S. J. (2008) assessed to physical activity (PA). Despite this, not enough is known about pupils' levels of physical activity or whether the delivery of these initiatives in schools facilitates or creates a barrier for pupils' PA. The aim of this study was to gain an understanding of adolescents PA levels from the perspective of those responsible for delivering physical education (PE) in schools; heads of PE (HOPE) and heads of school (HS). Seventeen semi-structured qualitative interviews were carried out with a snowball sample of HOPE and HS in schools in the Northwest and Southwest of England. Thematic data analysis using NVIVO was used to identify emergent themes. Despite many positive perceptions of the delivery of PE in schools, it is evident that barriers still exist within that delivery which discourages physical activity. More research is needed to particularly address the complex issues of elitism and the ethos of PA in schools.

Boyle, S. E., Jones, G. L., & Walters, S. J. (2008) assessed to physical activity (PA). Despite this, not enough is known about pupils' levels of physical activity or whether the delivery of these initiatives in schools facilitates or creates a barrier for pupils' PA. The aim of this study was to gain an understanding of adolescents PA levels from the perspective of those responsible for delivering physical education (PE) in schools; heads of PE (HOPE) and heads of school (HS). Seventeen semi-structured qualitative interviews were carried out with a snowball sample of HOPE and HS in schools in the Northwest and Southwest of England. Thematic data analysis using NVIVO was used to identify emergent themes. Despite many positive perceptions of the delivery of PE in schools, it is evident that barriers still exist within that delivery which discourages physical activity. More research is needed to particularly address the complex issues of elitism and the ethos of PA in schools.

Brooker, K., Van Dooren, K., McPherson, L., Lennox, N., & Ware, R. (2015) suggested that most adults with intellectual disability do not participate in sufficient amounts of physical activity (PA). A systematic review of peer-reviewed studies that reported an intervention aiming to improve PA levels of adults with intellectual disability was conducted.Keywords related to intellectual disability and physical activity were used to search relevant databases. Studies were excluded if they did not measure PA as an outcome for adults with intellectual disability, were non-English, and were not peer-reviewed. All relevant studies were included in the review regardless of methodological quality and design.PA interventions have the potential to improve the health and wellbeing of people with intellectual disability, a vulnerable group who require attention from public health practitioners and researchers. Given the health inequities that exist, public health researchers should target efforts to improve PA levels among this group.

Brown, S. A. (2005) evaluated the psychometric properties and relationship to physical activity levels of the Exercise Benefits/Barriers Scale (EBBS) among college students. A total of 398 college students completed the EBBS and a measure of self-efficacy, the Physical Exercise Self-Efficacy Scale. In addition, a subsample of 275 students also completed a semi structured interview on physical activity, the Seven-day Physical Activity Recall. Surprisingly, the factorial structure differed from the test developers'.

Brown, S. A., Huber, D., & Bergman, A. (2006) developed a new scale, the Physical Activity Benefits and Barriers Scale, and determine whether the scale was appropriate for college students. An exploratory factor analysis of 26 benefit and 24 barrier items of physical activity was administered to 493 college students, followed by an examination of the relationship between the newly derived factors and physical activity, as well as test-retest reliability. Psychometric properties were acceptable on this new scale, with the low motivation factor accounting for the highest amount of the variance in strenuous physical activity.

Brown, S. C., Geiselman, P. J., Copeland, A. L., Carol Gordon RN, P. H. D., & Richard-Eaglin, A. (2008) assessed The Personal Wellness Profile (PWP) is a 75-item questionnaire used to assess an individual's level of wellness in clinical and non-clinical settings. Currently, there are no published scientific data regarding the validity and reliability of the PWP. Therefore, this study was designed to assess convergence validity of the PWP using established instruments in which validity has been previously supported and to determine the reliability of the PWP using the test-retest method. A correlational design was used with a non-probability convenience sample of 100 African American women, ages 18 to 40 years, enrolled in a university in southern Louisiana. Convergence validity of the PWP was assessed by using the Michigan Alcoholism Screening Test (MAST), Drug Abuse Screening Test (DAST), Geiselman Food Preference Questionnaire II (GFPQ II), Weekly Stress Inventory (WSI), Fagerstrom Test for Nicotine Dependence (FTND), and the 7-Day Physical Activity Recall Questionnaire (PAR) as criteria. The PWP will be valuable as a valid, standardized, highly portable tool that can be taken into the community and administered to African-Americans and other populations who do not readily have access to treatment, research and health care centers.

Busse, J. W., Montori, V. M., Krasnik, C., Patelis-Siotis, I., & Guyatt, G. H. (2009) conducted a systematic review and meta-analysis to determine the efficacy of psychological interventions for premenstrual syndrome. We systematically searched and selected studies that enrolled women with premenstrual syndrome in which investigators randomly assigned them to a psychological intervention or to a control intervention. Trials were included irrespective of their outcomes and, when possible, we conducted meta-analyses. Low quality evidence from randomized trials suggests that cognitive behavioural therapy may have important beneficial effects in managing symptoms associated with premenstrual syndrome.

Dennerstein, L., Lehert, P., & Heinemann, K. (2012) reviewed published literature on the types and prevalences of premenstrual disorders and symptoms, and effects of these on activities of daily life and other parameters of burden of illness. The method involved review of the pertinent published literature. Premenstrual disorders vary in prevalence according to the definition or categorization. The most severe disorder being premenstrual dysphoric disorder (PMDD) affects 3-8% of women of reproductive age. This disorder focuses on psychological symptoms whereas global studies show that the most prevalent premenstrual symptoms are physical. Both psychological and physical symptoms affect women's activities of daily life. A considerable burden of illness has been shown to be associated with moderate to severe premenstrual disorders. In conclusion, premenstrual symptoms are a frequent source of concern to women during their reproductive lives and moderate to severe symptoms impact on their quality of lives.

Di Giulio, G., & Reissing, E. D. (2006) reviewed the current literature on PMDD focusing on prevalence, biopsychosocial etiological correlates, history of the development of a formal DSM diagnosis, and the controversies surrounding the current classification of PMDD. The authors conclude that PMDD presents a distinct diagnostic entity and that recognition through formal diagnostic criteria serves the important minority of women who suffer from this cyclical mood disorder distinct from premenstrual symptoms and major depression.

Fenton, C., Wellington, K., Moen, M. D., & Robinson, D. M. (2007) determined Drospirenone 3mg with ethinylestradiol 20microg (Yaz) is a low-dose combined oral contraceptive (COC) administered in a regimen of 24 days of active tablets followed by a short hormone-free interval (4 days; 24/4 regimen). Drospirenone, unlike other synthetic progestogens used in COCs, is a 17alpha-spirolactone derivative and a 17alpha-spironolactone analogue with antimineralocorticoid and antiandrogenic properties. Drospirenone/ethinylestradiol 3mg/20microg (24/4) is approved in the US for the prevention of pregnancy in women, for the treatment of the symptoms of premenstrual dysphoric disorder (PMDD) and for the treatment of moderate acne vulgaris in women who wish to use an oral contraceptive for contraception.Drospirenone/ethinylestradiol 3mg/20microg (24/4) provided 99% contraceptive protection over 1 year of treatment in two large studies. The same treatment regimen over three treatment cycles also significantly improved the emotional and physical symptoms associated with PMDD, and improved moderate acne vulgaris over six treatment cycles in double-blind trials. It was generally well tolerated, with adverse events generally typical of those experienced with other COCs and which were most likely to occur in the first few cycles. Clinical trials indicate that drospirenone/ethinylestradiol 3mg/20microg (24/4) is a good long-term contraceptive option, and additionally offers relief of symptoms that characterise PMDD and has a favourable effect on moderate acne vulgaris.

Ford, O., Lethaby, A., Roberts, H., & Mol, B. W. J. (2009) determined if progesterone has been found to be an effective treatment for all or some premenstrual symptoms and if adverse events associated with this treatment have been reported. We included randomised double-blind, placebo-controlled trials of progesterone on women with PMS diagnosed by at least two prospective cycles, without current psychiatric disorder. Two reviewers (BM and OF) extracted data independently and decided which trials to include. OF wrote to trial investigators for missing data. The trials did not show that progesterone is an effective treatment for PMS nor that it is not. Neither trial distinguished a subgroup of women who benefited, nor examined claimed success with high doses.

Ford, O., Lethaby, A., Roberts, H., & Mol, B. W. J. (2012) discussed Premenstrual Syndrome (PMS) is the term for severe symptoms experienced by about 5% of menstruating women up to two weeks before their menstrual periods, but not at other times. Treatment with progesterone may restore a deficiency, or balance the level of progesterone with other menstrual hormones. Progesterone therapy may reduce the effects of falling progesterone levels on the brain or on electrolytes in the blood. The objectives were to determine if progesterone has been found to be an effective treatment for all or some premenstrual symptoms, and if adverse events associated with this treatment have been reported. Two reviewers (BM and OF) extracted data independently, and decided on the trials to be included. OF wrote to the trial investigators to ask for missing data. We could not say that progesterone helped women with PMS, nor that it was ineffective. Neither trial distinguished a subgroup of women who benefited.

Foulds, H. J., Bredin, S. S., & Warburton, D. E. (2011) aimed to examine the health benefits of a community-based physical activity intervention for Aboriginal Canadians. From 2007–2010, 273 participants from the province of British Columbia, Canada were recruited through 21 Aboriginal communities representing male and females of wide ranging ages and health statuses. Participants attended identical testing of body composition, blood pressure, total and high density lipoprotein cholesterols, and physical activity behavior pre-and post-training. Participants completed one of three self-selected 13 week interventions: walking (n = 149), walk/running (n = 86), or running (n = 27).The self-selected intensities program was successful in improving health status and physical activity for Aboriginal adults of all ages, genders and activity programs.

Freeman, E. W. (2010) indicated that up to 25% of women may warrant treatment for the distress or impaired functioning associated with the symptoms.The primary focus of this review is the clinical condition of PMS and randomized, placebo-controlled trials of PMS treatment. A literature search in PubMed was conducted for these topics. The most recent reports of specific treatments in controlled treatment studies and all meta-analyses were selected.Reports consistently indicate that serotonergic antidepressants reduce PMS symptoms compared to a placebo. Hormonal treatments are the most widely prescribed medical treatment. Some oral contraceptives and gonadotropin-releasing hormone analogs have demonstrated efficacy, particularly for women who want contraception and PMS symptom control. Numerous non-pharmacologic treatments are utilized, but efficacy is reported only for calcium supplements, Vitex agnus castus (chasteberry), and cognitive-behavioral therapies. Further research to develop new therapies for the 40% of women with PMS who do not respond to the currently available treatments is needed.There are treatments with demonstrated efficacy for PMS, and the majority of women can be helped.

Jarvis, C. I., Lynch, A. M., & Morin, A. K. (2008) evaluated the current nonpharmacologic and pharmacologic treatment options for symptoms of premenstrual syndrome (PMS) and premenstrual dysphoric disorder (PMDD).Literature was obtained through searches of MEDLINE Ovid (1950-March week 3, 2008) and EMBASE Drugs and Pharmacology (all years), as well as a bibliographic review of articles identified by the searches. Key terms included premenstrual syndrome, premenstrual dysphoric disorder, PMS, PMDD, and treatment. All pertinent clinical trials, retrospective studies, and case reports in human subjects published in the English language were identified and evaluated for the safety and efficacy of pharmacologic and nonpharmacologic treatments of PMS/PMDD. Data from these studies and information from review articles were included in this review. Selective serotonin-reuptake inhibitors (SSRIs) have been proven safe and effective for the treatment of PMDD and are recommended as first-line agents when pharmacotherapy is warranted. Currently fluoxetine, controlled-release paroxetine, and sertraline are the only Food and Drug Administration-approved agents for this indication. Suppression of ovulation using hormonal therapies is an alternative approach to treating PMDD when SSRIs or second-line psychotropic agents are ineffective; however, adverse effects limit their use. Anxiolytics, spironolactone, and nonsteroidal antiinflammatory drugs can be used as supportive care to relieve symptoms. Despite lack of specific evidence, lifestyle modifications and exercise are first-line recommendations for all women with PMS/PMDD and may be all that is needed to treat mild-to-moderate symptoms. Herbal and vitamin supplementation and complementary and alternative medicine have been evaluated for use in PMS/PMDD and have produced unclear or conflicting results. More controlled clinical trials are needed to determine their safety and efficacy and potential for drug interactions. Healthcare providers need to be aware of the symptoms of PMS and PMDD and the treatment options available. Treatment selection should be based on individual patient symptoms, concomitant medical history, and need for contraception.

Jing, Z., Yang, X., Ismail, K. M., Chen, X. Y., & Wu, T. (2009) evaluated the effectiveness and safety of traditional Chinese herbal medicines in the treatment of women with premenstrual syndrome. Randomised controlled trials (RCTs) studying the efficacy of traditional Chinese herbal medicine(s) for treatment of the premenstrual syndrome were included.Two review authors telephoned the original authors of the RCTs to confirm the randomisation procedure, extracted and analysed data from the trials that met the inclusion criteria.It is rare in PMS management that efficacy claims are substantiated by clinical trials. One of the identified trials was well designed and reported on the effectiveness of Jingqianping in the treatment of premenstrual syndrome Qiao 2002. However, currently there is insufficient evidence to support the use of chinese herbal medicine for PMS and further, well controlled, trials are needed before any final conclusions could be drawn.

Kim, H. W., Kwon, M. K., Kim, N. S., & Reame, N. E. (2006) conducted to identify the potential relationship between the dietary intake level of soy isoflavones and perimenstrual symptomatology. The research design was a cross-sectional study. The sample was made up of 84 Korean women living in the USA, aged 28-40 years. The Moos Menstrual Distress Questionnaire (MDQ) and the Food Frequency Questionnaire were used as measurement tools. The soy isoflavone intake was significantly correlated with MDQ scores in the menstrual phase. In conclusion, the beneficial effect of dietary soy isoflavones on certain menstrual symptoms was established, suggesting that soy isoflavones could be one of the dietary factors related to the complexity of premenstrual syndrome (PMS). The positive effect of soy isoflavones on PMS warrants further study.

Kim, S. Y., Park, H. J., & Lee, H. (2011) evaluate the current evidence for acupuncture as a treatment for PMS.Ten databases were searched electronically, and relevant reviews were searched by hand through June 2009.Our review included randomised controlled trials (RCTs) of women with PMS; these RCTs compared acupuncture with sham acupuncture, medication, or no treatment. Study outcomes were presented as mean differences (for continuous data) or risk ratios (RRs) (for dichotomous data) with a 95% confidence interval (95% CI). The risk of bias was assessed using the assessment tool from the Cochrane Handbook. Although acupuncture seems promising for symptom improvement in women with PMS, important methodological flaws in the included studies weaken the evidence. Considering the potential of acupuncture, further rigorous studies are needed.

Kimura, N., Fukuwatari, T., Sasaki, R., Hayakawa, F., & Shibata, K. (2003) carried out a survey of vitamin intake in Japanese women who were university seniors majoring a dietitian course. The subjects (n = 33) consumed self-selected foods, and food intake was recorded by the weight method. We calculated the vitamin intake except for biotin from the food records using the Standard Food Tables of Japanese Foods. In terms of daily intake, vitamin A was 705+/-435 microg (mean+/-SD), vitamin D 6+/-8 microg, vitamin E 7.7+/-3.0 mg, vitamin K 191+/-156 microg, vitamin B1 0.7+/-0.3 mg (0.43+/-0.15 mg/1,000 kcal), vitamin B2 1.1+/-0.4 mg (0.65+/-0.18 mg/1,000 kcal), vitamin B6 0.9+/-0.4 mg (0.017+/-0.005 mg/g protein), vitamin B12 4.4+/-4.1 microg, niacin equivalent 23+/-7 mg (14.4+/-4.9 mg/1,000 kcal), pantothenic acid

4.6+/-1.4 mg, folic acid 267+/-115 microg, and vitamin C 73+/-38 mg. All of these averages were around the Japanese Recommended Dietary Allowance (RDA) for level "III (preferable)" of physical activity. Major vitamin A resources were vegetables; vitamin D resources, fish; vitamin E resources, fats and oils and vegetables; vitamin K resources, vegetables; vitamin B1 resources, cereals and animal meats; vitamin B2 resources, various foods; vitamin B6 resources, cereals, vegetables, fish, and animal meats; vitamin B12 resources, fish; niacin equivalent resources, fish, animal meats, and cereals; pantothenic acid resources, various foods; folic acid resources, vegetables; and vitamin C resources, vegetables and potatoes. From this survey, it was found that Japanese women college students consumed many kinds of food, and therefore, their vitamin nutrition was good as compared to the RDA values for level III of physical activity; however, their energy intake (1,622+/-377 kcal) was lower than the RDA for level III (2,050 kcal/d). Their strength of physical activity would be level I. Therefore, in consideration of their lifestyle, their energy intakes is considered adequate. In conclusion, a problem for student lifestyle is a shortage of food intake due to lack of exercise.

Kleinstäuber, M., Witthöft, M., & Hiller, W. (2012) investigated the efficacy of psychotherapeutic interventions and psychopharmacotherapy for premenstrual syndrome (PMS) and premenstrual dysphoric disorder. Based on a multiple-phase literature search, controlled trials were selected according to a priori defined inclusion criteria. Data were extracted on the basis of a standardized coding scheme. The standardized weighted mean difference (random effects model) was used as effect size index. Dependent on outcome, 22 included studies obtained small to medium effect sizes for cognitive-behavioral interventions (range: d(+) = 0.24-0.70) and for serotonergic antidepressants (range: d(+) = 0.29-0.58), at post-assessment. Follow-ups were performed only in studies of cognitive-behavioral interventions (range: d(+) = 0.46-0.74). There was no evidence of a publication bias. For both cognitive-behavioral interventions and serotonergic antidepressants, efficacy in treatment of PMS was found to not be satisfactory. Future research should possibly focus more on a combination of both approaches.

Koltyn, K. F. (2001) examined the association between physical activity and quality of life in older women. Women over the age of 60 years, living independently or in assisted-care facilities, answered a number of questions pertaining to their health status, physical activity levels, and quality of life. Results indicated that the women living independently had significantly ($P < .05$) higher physical activity levels compared to the women living in assisted-care facilities. In addition, overall quality of life and the domains of physical health, social relationships, and environment were found to be significantly ($P < .05$) higher in the women living independently compared to the women living in assisted-care facilities. Correlational analyses revealed that physical activity levels correlated significantly ($P < .05$) with overall quality of life and the physical health domain. It is concluded that there is an association between physical activity levels and quality of life in older women.

Kondo, T., Kobayashi, I., & Murakami, M. (2006) studied the effect of exercise on circulating adipokine, high sensitivity C-reactive protein (hs-CRP), and metabolic parameters in obese young women. Ninety-six healthy Japanese young female students aged 18-23 years were studied. The longitudinal intervention study of a 7-month exercise training program (30-60 min/day, 60-70% HR-reserve, 200-400 kcal, 4-5 days/week) was performed in eight obese female students (BMI > or =25 kg/m(2)). Eight control female students (mean BMI = 22 kg/m(2)) were included in the follow-up study. Body weight, body mass index (BMI), percentage of body fat (%Fat), body fat mass, lean body mass, health-promoting lifestyle profile-scale (L-scale), VO(2)max (maximal oxygen uptake), hs-CRP, lipids, insulin homeostasis model assessment (HOMA-R), fasting levels for circulating adiponectin, leptin, and TNF-alpha, were measured before and after the exercise program. In obese subjects, body weight, BMI, %Fat, body fat mass, lean body mass, hs-CRP, leptin, and TNF-alpha were significantly higher, and L-scale and adiponectin were lower than those in control subjects. In obese subjects, exercise decreased body weight, BMI, %Fat, body fat mass, lean body mass, hs-CRP, leptin, and TNF-alpha, and increased L-scale, VO(2)max, HDL-cho, and adiponectin. It was concluded that changes in circulating adipokine levels are involved in the improvement of the metabolic state by exercise and may be useful markers for evaluation and prescription of exercise.

Kuhl. (2002) discussed Estradiol, progesterone and some of their metabolites modulate the activity of neurotransmitters and neuropeptides in the

CNS. The distribution and concentrations of sex steroids in the various CNS regions is partly dependent on the serum levels, but also on the local synthesis of the steroids. In general, estradiol and testosterone exert a stimulatory, progesterone an inhibitory effect on neuronal activities which are mediated by excitatory (e.g. glutamate, aspartate), and inhibitory amino acids (e.g. GABA) and neuropeptides (e.g. beta-endorphin), respectively. Gonadotropin release is primarily governed by the rhythm of pulsatile secretion of GnRH in the hypothalamus which is controlled by estradiol and progesterone by means of inhibitory or stimulatory modulation of the amplitude and frequency of GnRH pulses. The discharges of GnRH neurons triggered by excitatory amino acids are modulated by estradiol, while the inhibitory effect of progesterone is mediated by GABA and beta-endorphin which cause hyperpolarization of the GnRH neurons and consequently a reduced pulse frequency. The pulse amplitudes are primarily influenced by estradiol, but neuropeptide Y, neurotensin and noradrenaline contribute to their preovulatory enhancement. The postovulatory rise in core temperature is caused by the increasing level of progesterone and its metabolite 3 alpha-pregnanolone, respectively. Despite of this, up to 20% of ovulatory cycles do not show any rise in body temperature. Although 3 alpha-pregnanolone has sedative activities, there is no change in sleep quality during the luteal phase due to their low serum levels. It could be demonstrated that performance on tests of articulatory and fine motor skills are enhanced in the late follicular phase as compared to the menstruation phase, while spatial ability was better during menses. Estrogens may influence mood and well-being in a favorable manner, while in predisposed women progesterone may cause symptoms of premenstrual syndrome. In most women there are, however, no cycle-dependent mood changes. An increase in appetite can be observed during the periovulatory phase and before menses, while sexual interest increases in the follicular phase. Somatic complaints (back pain, abdominal pain, breast tenderness) which are highest before and during menstruation, are probably associated with a lowered pain threshold due to a fall in the beta-endorphin levels in the CNS.

Kulavic, K., Hultquist, C. N., & McLester, J. R. (2013) investigated the motivational factors and the barriers to physical activity (PA) in traditional college students (TS) and nontraditional college students (NTS) and determine if differences exist between these 2 groups. A total of 746 college students; 628 were TS (19.1 ± 1.2 years), and 118 were NTS (31.2 ± 8.6 years). In fall 2010, students voluntarily completed a demographic questionnaire, Barriers to Being Active Quiz, and Exercise Motivations Inventory-2 (EMI-2). Based on these findings, interventions can be developed for the varied needs of a diverse student population as it pertains to PA participation.

Kwan, I., & Onwude, J. L. (2009) discussed Premenstrual symptoms occur in 95% of women of reproductive age. Severe, debilitating symptoms (PMS) occur in about 5% of those women. There is no consensus on how symptom severity should be assessed, which has led to a wide variety of symptoms scales, making it difficult to synthesise data on treatment efficacy. The cyclical nature of the condition also makes it difficult to conduct RCTs.

We conducted a systematic review and aimed to answer the following clinical questions: What are the effects of drug treatments; hormonal treatments; psychological interventions; physical therapies; dietary supplements; and surgical treatments in women with premenstrual syndrome? We searched: Medline, Embase, The Cochrane Library and other important databases up to November 2006 (Clinical Evidence reviews are updated periodically, please check our website for the most up-to-date version of this review). We included harms alerts from relevant organisations such as the US Food and Drug Administration (FDA) and the UK Medicines and Healthcare products Regulatory Agency (MHRA).In this systematic review we present information relating to the effectiveness and safety of the following interventions: acupuncture, alprazolam, bright light therapy, buspirone, chiropractic manipulation, clomipramine, cognitive behavioural therapy, danazol, endometrial ablation, evening primrose oil, exercise, gonadorelin analogues, hysterectomy, laparoscopic bilateral oophorectomy, magnesium supplements, metolazone, non-steroidal anti-inflammatory drugs, oestrogens, oral contraceptives, progesterone, progestogens, pyridoxine, reflexology, relaxation, selective serotonin reuptake inhibitors, spironolactone, tibolone.

Madanat, H., & Merrill, R. M. (2006) investigated physical activity levels across the five stages of change for physical activity and to identify motivational factors for physical activity according to these stages of change among college students in Amman, Jordan. Analyses were based on a cross-sectional survey of 431 students, with a mean age of 21.1 (SD=0.16) and

67.5% female. Based on the recommendation that physical activity requires at least 30 minutes of physical activity 3 or more days per week, men were more likely than women to classify themselves in later stages: 7.3% vs. 9.5% in the precontemplation stage, 17.4% vs. 14.7% in the contemplation stage, 50.0% vs. 63.5% in the preparation stage, 9.4% vs. 5.6% in the action stage, and 15.9% vs. 6.7% in the maintenance stage [X2(4) = 14.04, p = 0.0072]. Seven potential motivational items for physical activity were assessed using factor analysis: experience better self-worth, prevent chronic disease, relieve stress, stay in shape, longevity, recreation/fun, and social benefits. Two factor groupings were identified from these items. The first factor included the first five items, labeled as "Physical and Mental". The second factor included the last two items, labeled as "Social and Recreational." "Physical and Mental" items compared with "Social and Recreational" items were most likely to motivate physical activity across the stages of change for physical activity. The strongest motivator of physical activity was to stay in shape. The weakest motivator of physical activity was for social reasons. The influence of the intermediate motivational factors was slightly affected by the students' stage of change for physical activity. Motivators for physical activity did not differ according to sex. These results provide important information about the motivational factors for physical activity for college-aged students in Jordan that can be useful in developing effective physical activity intervention programs.

Maddison, R., Pfaeffli, L., Whittaker, R., Stewart, R., Kerr, A., Jiang, Y., & Rawstorn, J. (2015) determined the effectiveness and cost-effectiveness of a mobile phone intervention to improve exercise capacity and physical activity behaviour in people with ischaemic heart disease (IHD).A mobile phone intervention was not effective at increasing exercise capacity over and above usual care. The intervention was effective and probably cost-effective for increasing physical activity and may have the potential to augment existing cardiac rehabilitation services.

Makuch, M. Y., Osis, M. J., de Pádua, K. S., & Bahamondes, L. (2013) assessed the understanding and attitudes of Brazilian men regarding premenstrual syndrome (PMS). In a survey-based study between September 2007 and April 2008, information was collected from men aged 18-40 years who were attending public healthcare services or were university and faculty staff at 5 cities in different geographic regions of Brazil and the Federal District. Many of the men interviewed were knowledgeable about PMS symptoms; however, this awareness was more common among men of higher socioeconomic strata with more years of schooling.

Malinauskas, B. M., Raedeke, T. D., Aeby, V. G., Smith, J. L., & Dallas, M. B. (2006) concerned to health educators is the suggestion that college females practice diet and health behaviors that contradict the 2005 dietary guidelines for Americans. In this regard, there remain gaps in the research related to dieting among college females. Namely, do normal weight individuals diet differently from those who are overweight or obese, and are there dieting practices used by females that can be adapted to promote a healthy body weight? Since it is well recognized that females diet, this study seeks to determine the dieting practices used among normal, overweight, and obese college females (do they diet differently) and identify dieting practices that could be pursued to help these females more appropriately achieve and maintain a healthy body weight. A total of 185 female college students aged 18 to 24 years participated in this study. Height, weight, waist and hip circumferences, and skinfold thickness were measured to assess body composition. Surveys included a dieting practices questionnaire and a 30-day physical activity recall. Participants were classified according to body mass index (BMI) as normal weight (n = 113), overweight (n = 35), or obese (n = 21). Data were analyzed using JMP IN(R) software. Descriptive statistics included means, standard deviations, and frequency. Subsequent data analysis involved Pearson X2 and one-way analysis of variance with comparison for all pairs that were significantly different using Tukey-Kramer honestly significant difference test. Collectively, results indicate female college students, regardless of weight status, would benefit from open discussions with health educators regarding healthy and effective dieting practices to achieve/maintain a healthy body weight. The results are subject to replication among high school, middle-aged, and older females.

Martens, M. P., Buscemi, J., Smith, A. E., & Murphy, J. G. (2012) examined the short-term efficacy of a brief motivational intervention (BMI) designed to increase physical activity.Participants were 70 college students who reported low physical activity (83% women, 60% African American). Participants were randomly assigned to either the BMI condition or to an education-only (EO) condition. They completed measures of physical activity at baseline and 1-month follow-up.The findings from this study provide preliminary support for the efficacy of a BMI designed to increase physical activity among college students. Future studies should continue to examine and refine the intervention in an effort to improve health-related behaviors among this group.

Matsudo, V., Matsudo, S., Andrade, D., Araujo, T., Andrade, E., de Oliveira, L. C., & Braggion, G. (2002) presented key points of an intervention programme (Agita São Paulo Program) to promote physical activity in a developing country. Agita is a multi-level, community-wide intervention designed to increase knowledge about the benefits and the level of physical activity in a mega-population of 34 million inhabitants of São Paulo State, Brazil. The main message was taken from the Centers for Disease Control/American College of Sports Medicine (CDC/ACSM) recommendation that: 'everyone should accumulate at least 30 minutes of physical activity, on most days of the weeks, of moderate intensity, in one single or in multiple sessions'. Activities were encouraged in three settings: home, transport and leisure time. Focus groups were students from elementary schools through to college, white and blue collar workers, and elderly people. Innovative aspects included: (1) a research centre leading the process, (2) scientific and institutional partnerships (over 160 groups), (3) a feasible approach--the 'one-step-ahead' model, (4) empowerment, (5) inclusion, (6) non-paid media, (7) social marketing, and (8) culture-linked. Data were obtained from 645 random, home-based questionnaires over four years--stratified by sex, age, education and socio-economic level. These data show that the Agita message reached 55.7% of the population, and among these, 23.1% knew the main message. Recall of Agita and knowledge of its purpose were well distributed among different socioeconomic levels, being known by 67% of the most educated. The prevalence of people reaching the recommendation was 54.8% (men 48.7%, women 61%); and risk of being sedentary was quite smaller among those who knew the Agita message (7.1%) compared with those who did not know (13.1%). In conclusion, based upon the Agita São Paulo experience, it appears that a multi-level, community-wide intervention to promote physical activity may obtain good results if the model contains the items listed above.

Matsumoto, T., Asakura, H., & Hayashi, T. (2013) discusses a majority of women in their reproductive years experience a variety of symptoms premenstrually that can alter behavior and well-being and affect family, friends, and working relationships. Notwithstanding its prevalence, however, research has not yet clarified this inscrutable condition, commonly known as premenstrual syndrome (PMS) or more severe PMS, premenstrual dysphoric disorder (PMDD). This comprehensive review discusses the diagnosis, epidemiology, symptoms, etiology, and the complex web of biopsychosocial factors that attends PMS.

Mazlyn Mena, M., Nagarajah Lee, H. L., Fatimah, A., Norimah, A. K., & Goh, K. L. (2013) aimed to elucidate the relations between stool patterns, dietary intake and physical activity levels among adults with functional constipation.From a database collected via surveys at public events, a convenience sample of 100 adults diagnosed with Rome II-defined functional constipation was enrolled in this cross-sectional study. After severity assessment using the Chinese Constipation Questionnaire, subjects completed 2-week bowel movement diaries to determine stool frequency, consistency and output. Dietary intake and physical activity levels were assessed twice using three-day 24-hour diet recalls and International Physical Activity Questionnaire, respectively. Ninety subjects who completed the study were included in the analysis.Insufficient intake of dietary fibre and fluid are associated with aggravated constipation symptoms. Further studies are necessary to confirm usefulness of dietary intervention in treatment of constipation as dietary factors alone may not influence overall severity and stool consistency, an integral element of constipation.

McKercher, C., Schmidt, M. D., Sanderson, K., Dwyer, T., & Venn, A. J. (2012) demonstrated physical activity domain specificity in associations between physical activity and depression in young adults participating in the Childhood Determinants of Adult Health study. The current analysis examined associations between physical activity in different domains (active commuting, school physical education, discretionary sport) and depressed mood when cohort participants were children and adolescents.Participants were 6070 school-children (aged 9–15 years) participating in a national health and fitness survey in 1985. Physical activity and depressed mood were assessed via self-report. Analyses were conducted separately by gender and school-level, and prevalence ratios adjusted for scholastic level, health status, smoking, alcohol consumption, residential arrangements and familial language.Associations between physical activity and depressed mood varied by gender, school-level and the domain in which activity was assessed. Further research utilizing more accurate assessment of childhood physical activity and depression is required to clarify potential associations.

Meersseman, W., Cassiman, D., Goossens, W., & Vanderschueren, S. (2008) analyzed 22-year-old woman was admitted to intensive care with severe hyponatraemia. She suffered from lower abdominal pain, vomiting and irritability since one week. Physical findings showed euvolemia and an altered mental status with severe agitation and slurred speech. Abdominal examination was painful but there were no signs of peritonitis. Laboratory data were compatible with the diagnosis of syndrome of inappropriate secretion of antidiuretic hormone. Since patient was in a premenstrual phase, recently started to take an oral contraceptive and since no abnormalities were seen on an abdominal CT scan, the presentation was considered suggestive of an acute porphyria attack. A urinary sample indicated markedly increased levels of delta-aminolevulinic acid, porphobilinogen and uroporphyrin. A low activity of the porphobilinogen deaminase enzyme confirmed the diagnosis of acute intermittent porphyria. The present case demonstrates the need for a high level of suspicion in order to diagnose this disorder in unexplained syndrome of inappropriate antidiuretic hormone secretion and prevent life-threatening complications.

Merom, D., Delbaere, K., Cumming, R., Voukelatos, A., Rissel, C., Van der Ploeg, H. P., & Lord, S. R. (2014) assessed the criterion validity and responsiveness of the Incidental and Planned Exercise Questionnaire (IPEQ) specifically developed for aging research.The sample comprised 315 community-dwelling inactive older adults (mean age = 73.2 yr) who participated in a trial investigating the effect of a walking program on falls. At baseline, a subsample (n = 177) also wore an accelerometer (ActiGraph GT1M) for 7 d, and 126 of those had four valid days or more of accelerometer data. Validity coefficients (Spearman ρ) were calculated between accelerometer counts per minute, average steps per day, average moderate- to vigorous-intensity physical activity (MVPA) using two cut points (760 and 1041 counts per minute), and IPEQ-MVPA (the sum of physical activity hours per week derived from nine IPEQ items excluding the indoor chores question). Responsiveness was the mean IPEQ change in the intervention group divided by the SD of the mean change in control group.IPEQ provides a practical and valid measure of MVPA in surveillance and intervention studies. Incidental activities are mostly of a light-intensity nature, and their low ambulatory component may explain the lower correlation between IPEQ as a whole and accelerometer measures.

Mosallanezhad, Z., Salavati, M., Sotoudeh, G. R., Wikmar, L. N., & Frändin, K. (2014) compared elderly people who walked at least 30 min a day with others who walked less, from a gender perspective, regarding perceived health and fitness, physiological capacity and functional performance. A representative sample of 75-year-olds born 1932–33 and living in Tehran, in 2007–2008 was included by randomly selecting 1100 subjects from the latest Iranian census records (1996) by the Statistical Centre of Iran using computerized methods. Participants answered questions regarding health status and physical activity and performed functional tests. Better results for Walkers were observed in most subjective and objective outcome measures. Walkers were less likely to feel generally tired, more likely to have better physical fitness and to have the maximum score on the Falls Efficacy Scale, less likely to feel unstable during walking outdoors and less likely to be dependent or unsafe in ADL. Walkers of both genders performed better in the following tests: chair stand, one leg stance, maximal walking speed and six min walking. The difference between Walkers and Non-Walkers was greater in men. In general, older women and men who walked at least 30 min daily/almost daily showed better results in most health-related outcomes, ADL and functional performance than people who walked less. This study showed gender differences in the level of physical activity and functioning that must be taken into account when planning intervention programs.

Myers, R. S., & Roth, D. L. (1997) analyzed college students (N = 432) completed questionnaires assessing current exercise participation, intention to become more active, perceived benefits of exercise, and perceived barriers to exercise. Confirmatory factor analyses of the benefit and barrier items suggested that a model containing 4 benefit factors (social, psychological, body image, and health) and 4 barrier factors (time-effort, social, physical effects, and specific obstacles) fit better than other models with fewer factors. Subscales derived from these factors revealed specific differences across participants who were categorized into 1 of 4 stages of exercise adoption. These findings confirm the multidimensional nature of perceived benefits of and barriers to exercise. Implications for the promotion and maintenance of exercise are discussed.

Nelson, M. C., Lust, K., Story, M., & Ehlinger, E. (2008) examined cross-sectional associations between credit card debt, stress, and health risk behaviors among college students, focusing particularly on weight-related behaviors. Random-sample, mailed survey. Undergraduate and graduate students (n = 3206) attending a large public university. Self-reported health indicators (e.g., weight, height, physical activity, diet, weight control, stress, credit card debt).University student lifestyles may be characterized by a variety of coexisting risk factors. These findings indicate that both debt and stress were associated with wide-ranging adverse health indicators. Intervention strategies targeting at-risk student populations need to be tailored to work within the context of the many challenges of college life, which may serve as barriers to healthy lifestyles. Increased health promotion efforts targeting stress, financial management, and weight-related health behaviors may be needed to enhance wellness among young adults.

Nickelson, J., Alfonso, M. L., McDermott, R. J., Bumpus, E. C., Bryant, C. A., & Baldwin, J. A. (2011) discussed about creating community-based opportunities for youth to be physically active is challenging for many municipalities. A Lexington, Kentucky community coalition designed and piloted a physical activity program, 'VERB™ summer scorecard (VSS)', leveraging the brand equity of the national VERB™--It's What You Do! campaign. Key elements of VSS subsequently were adopted in Sarasota County, FL. This study identified characteristics of Sarasota's VSS participants and non-participants. Students in Grades 5-8 from six randomly selected public schools completed a survey assessing VSS participation, physical activity level, psychosocial variables, parental support for physical activity and demographics. Logistic regression showed that VSS participants were more likely to be from Grades 5 to 6 versus Grades 7 and 8 [odds ratio (OR) = 6.055] and perceive high versus low parental support for physical activity (OR = 4.627). Moreover, for each unit rise in self-efficacy, the odds of VSS participation rose by 1.839. Chi-squared automatic interaction detector (CHAID) analysis suggested an interaction effect between grade and school socioeconomic status (SES), with a large proportion of seventh and eighth graders from high SES schools being non-participants (76.6%). A VSS-style program can be expected to be more effective with tweens who are younger, in a middle SES school, having high self-efficacy and high parental support for physical activity.

O'Brien, P. M. S., Bäckström, T., Brown, C., Dennerstein, L., Endicott, J., Epperson, C. N. & Panay, N. (2011) discussed Premenstrual disorders (PMD) are characterised by a cluster of somatic and psychological symptoms of varying severity that occur during the luteal phase of the menstrual cycle and resolve during menses (Freeman and Sondheimer, Prim Care Companion J Clin Psychiatry 5:30-39, 2003; Halbreich, Gynecol Endocrinol 19:320-334, 2004). Although PMD have been widely recognised for many decades, their precise cause is still unknown and there are no definitive, universally accepted diagnostic criteria. To consider this issue, an international multidisciplinary group of experts met at a face-to-face consensus meeting to review current definitions and diagnostic criteria for PMD. This was followed by extensive correspondence. The consensus group formally became established as the International Society for Premenstrual Disorders (ISPMD). The inaugural meeting of the ISPMD was held in Montreal in September 2008. The primary aim was to provide a unified approach for the diagnostic criteria of PMD, their quantification and guidelines on clinical trial design. This report summarises their recommendations. It is hoped that the criteria proposed here will inform discussions of the next edition of the World Health Organisation's International Classification of Diseases (ICD-11), and the American Psychiatric Association's Diagnostic and Statistical Manual of Mental Disorders, fifth edition (DSM-V) criteria that are currently under consideration. It is also hoped that the proposed definitions and guidelines could be used by all clinicians and investigators to provide a consistent approach to the diagnosis and treatment of PMD and to aid scientific and clinical research in this field.

Osorio, E. A., Raygoza, N. P., & Paloalto, M. L. R. (2013) compared relationship between perceived barriers and physical activity level in older adults from Aguascalientes in Mexico. The present study is cross-sectional, analytic, and observational.150 older adults registered in the Family Integral Development System from Aguascalientes State were included at random, who accepted to participate by signing a consent form. A questionnaire to detect perceived barriers as well as an international physical activity questionnaire was applied. Z for two proportions, p-value and odds ratio, 95% confidence interval were calculated, between external and internal perceived barriers and low physical activity level in older adults from the sample. The nursing designing programs to perform physical activity in older adults should take into account these perceived barriers.

Ossewaarde, L., Van Wingen, G. A., Kooijman, S. C., Bäckström, T., Fernández, G., & Hermans, E. J. (2010) suggested that such effects may result from actions of gonadal hormones on the mesolimbic dopamine (DA) system. We therefore investigated premenstrual changes in reward-related neural activity in terminal regions of the DA system in humans. Twenty-eight healthy young women underwent functional magnetic resonance imaging on 2 days during the menstrual cycle, once during the late follicular phase and once during the premenstrual phase, in counterbalanced order. Using a modified version of the monetary incentive delay task, we assessed responsiveness of the ventral striatum to reward anticipation. Our results show enhanced ventral striatal responses during the premenstrual as compared to the follicular phase. Moreover, this effect was most pronounced in women reporting more premenstrual symptoms. These findings provide support for the notion that changes in functioning of mesolimbic incentive processing circuits may underlie premenstrual changes in motivated behaviors. Notably, increases in reward-cue responsiveness have previously been associated with DA withdrawal states. Our findings therefore suggest that the sharp decline of gonadal hormone levels in the premenstrual phase may trigger a similar withdrawal-like state.

Pereira-Vega, A., Sánchez, J. L., Gil, F. L., Maldonado, J. A., Bravo, J. M., Ignacio, J. M., ... & Sánchez, I. (2010) compared premenstrual symptoms among asthmatic women according to whether they have premenstrual asthma or not.A questionnaire on respiratory symptoms during a single menstrual cycle was developed for asthmatics of fertile age, together with another on symptoms related to premenstrual syndrome. These included dysphoric-psychiatric symptoms (anxiety, depression, fatigue, irritability, and mood swings), edematous symptoms (abdominal and mammary tension, swelling, acne, and migraine), and other symptoms (leg pains, nausea, sweating, vomiting, and tiredness). Morning and evening peak flow scores were collected to evaluate lung function. Premenstrual asthma was determined to be a $\geq$ 20% objective exacerbation in the premenstrual phase over the preovulatory phase in terms of both respiratory symptoms and/or peak flow. The intensity of the change in symptoms was evaluated according to the effect size.A clear link was found between premenstrual asthma and the premenstrual exacerbation of dysphoric symptoms, and certain edematous symptoms such as abdominal and mammary tension as well as a swelling sensation.

Phillips, M. M., Goodell, M., Raczynski, J. M., & Philyaw Perez, A. G. (2012) expressed epidemic increases in childhood obesity and associated health risks are resulting in efforts to implement school policies related to nutrition and physical activity (NPA). With multicomponent policy efforts, challenges exist in characterizing the extent of policy change across the breadth of NPA policies.Aggregated policy indices were created to characterize NPA policy implementation in Arkansas public schools from 2004 through 2009. Index scores are presented by year, domain, and school level.The policy index approach facilitates the consideration of the effect of school policy change in a holistic, aggregated way. School characteristics influence policy adoption, and thus, should be taken into consideration in the promotion of policy change.

Pramanik, T., & Pramanik, S. (2001) determined physical fitness of Nepalese students at a medical college in Kathmandu was measured. The proportions of students with poor fitness, average fitness, good fitness, very good fitness and excellent fitness were 10.4%, 23.6%, 39.6%, 17.9% and 8.5% respectively. No significant difference between the male and female students in physical fitness index score was noticed in any of the fitness level groups.

Ramírez-Vélez, R., Tordecilla-Sanders, A., Laverde, D., Hernández-Novoa, J. G., Ríos, M., Rubio, F., ... & Martínez-Torres, J. (2015) investigated the prevalence of barriers and their association with Colombia college students engaging in PA. A total of 5,663 students (3,348 male) from three cities in Colombia. In fall 2013, students voluntarily completed a demographic questionnaire, Barriers to Being Active Quiz. Logistic regression analysis of each barrier (adjusted for confusion variables: gender, age and BMI) was used for verifying such association. A significant prevalence was found regarding self-perception of barriers leading to students ceasing to engage in PA.

Rapkin, A. (2003) severed premenstrual syndrome (PMS) and, more recently, premenstrual dysphoric disorder (PMDD) have been studied extensively over the last 20 years. The defining criteria for diagnosis of the disorders according to the American College of Obstetricians and Gynecologists (ACOG) include at least one moderate to severe mood symptom and one physical symptom for the diagnosis of PMS and by DSM IV criteria a total of 5 symptoms with 1 severe mood symptom for the diagnosis of PMDD. There must be functional impairment attributed to the symptoms. The symptoms must be present for one to two weeks premenstrually with relief by day 4 of menses and should be documented prospectively for at least two cycles using a daily rating form. Nonpharmacologic management with some evidence for efficacy include cognitive behavioral relaxation therapy, aerobic exercise, as well as calcium, magnesium, vitamin B(6) L-tryptophan supplementation or a complex carbohydrate drink. Pharmacologic management with at least ten randomized controlled trials to support efficacy include selective serotonin reuptake inhibitors administered daily or premenstrually and serotonergic tricyclic antidepressants. Anxiolytics and potassium sparing diuretics have demonstrated mixed results in the literature. Hormonal therapy is geared towards producing anovulation. There is good clinical evidence for GnRH analogs with addback hormonal therapy, danocrine, and estradiol implants or patches with progestin to protect the endometrium. Oral contraceptive pills prevent ovulation and should be effective for the treatment of PMS/PMDD. However, limited evidence does not support efficacy for oral contraceptive agents containing progestins derived from 19-nortestosterone. The combination of the estrogen and progestin may produce symptoms similar to PMS, such as water retention and irritability. There is preliminary evidence that a new oral contraceptive pill containing low-dose estrogen and the progestin drospirenone, a spironolactone analog, instead of a 19-nortestosterone derivative can reduce symptoms of water retention and other side effects related to estrogen excess. The studies are in progress, however, preliminary evidence suggests that the drospirenone-containing pill called Yasmin may be effective the treatment of PMDD.

Rapkin, A. J., & Akopians, A. L. (2012) described about premenstrual syndrome (PMS) and premenstrual dysphoric disorder are triggered by hormonal events ensuing after ovulation. The symptoms can begin in the early, mid or late luteal phase and are not associated with defined concentrations of any specific gonadal or non-gonadal hormone. Although evidence for a hormonal abnormality has not been established, the symptoms of the premenstrual disorders are related to the production of progesterone by the ovary. The two best-studied and relevant neurotransmitter systems implicated in the genesis of the symptoms are the GABArgic and the serotonergic systems. Metabolites of progesterone formed by the corpus luteum of the ovary and in the brain bind to a neurosteroid-binding site on the membrane of the gamma-aminobutyric acid (GABA) receptor, changing its configuration, rendering it resistant to further activation and finally decreasing central GABA-mediated inhibition. By a similar mechanism, the progestogens in some hormonal contraceptives are also thought to adversely affect the GABAergic system. The lowering of serotonin can give rise to PMS-like symptoms and serotonergic functioning seems to be deficient by some methods of estimating serotonergic activity in the brain; agents that augment serotonin are efficacious and are as effective even if administered only in the luteal phase. However, similar to the affective disorders, PMS is ultimately not likely to be related to the dysregulation of individual neurotransmitters. Brain imaging studies have begun to shed light on the complex brain circuitry underlying affect and behaviour and may help to explicate the intricate neurophysiological foundation of the syndrome.

Rapkin, A. J., & Mikacich, J. A. (2008) reviewed the current knowledge about the prevalence, diagnosis, and management of premenstrual syndromes in adolescents.Large epidemiologic studies addressing adolescent premenstrual disorders, clinical presentation, and comorbidity with other disorders have yet to be performed. Randomized controlled treatment trials for teens with moderate-to-severe premenstrual syndrome or the more severe affective predominant, premenstrual dysphoric disorder still are sorely lacking. This review will present an updated review of the published studies with respect to premenstrual syndrome and premenstrual dysphoric disorder in adolescents in the context of the large body of literature regarding presentation, diagnosis, and treatment in adult women. Premenstrual disorders likely start in the teen years. At least 20% of adolescents may experience moderate-to-severe premenstrual symptoms associated with functional impairment. Current treatment includes lifestyle recommendations and pharmacologic agents that suppress the rise and fall of ovarian steroids or augment serotonin.

Rapkin, A. J., & Winer, S. A. (2008) explained premenstrual dysphoric disorder (PMDD) is characterized by physical, affective and behavioral symptoms that are linked to the luteal phase of the menstrual cycle and relieved soon after the onset of menses. The disorder is chronic and exerts a major impact on personal relationships and occupational productivity for the estimated 6% of reproductive-aged women who fulfill strict PMDD criteria and the almost 20% of women who nearly meet these criteria. There are now various pharmacologic options that have demonstrated efficacy for PMDD and two of these approaches have an approved indication for treatment from the US FDA: three selective serotonin re-uptake inhibitors; and for women who also desire hormonal contraception, a low dose oral contraceptive pill containing the progestin drospirenone, in a new dosing regimen. Due to the unique pathophysiology of the disorder, the selective serotonin re-uptake inhibitors can be effectively administered intermittently, with dosing limited to the luteal phase of the cycle (2 weeks prior to menses). In the future, new pharmacotherapy will likely evolve from research evaluating other hormonal formulations that inhibit ovulation, without simulating PMDD-like symptoms, or novel pharmacologic agents that modulate the central neurotransmission.

Rapkin, A. J., & Winer, S. A. (2009) described premenstrual symptoms are distressing for up to 20% of reproductive-aged women and are associated with impairment in interpersonal or workplace functioning for at least 3-8%. Typical symptoms of premenstrual syndrome and the severe form, premenstrual dysphoric disorder, include irritability, anger, mood swings, depression, tension/anxiety, abdominal bloating, breast pain and fatigue. The symptoms recur monthly and last for an average of 6 days per month for the majority of the reproductive years. For women with premenstrual dysphoric disorder, the symptoms can be as disabling as major depressive disorder. It has been estimated that affected women experience almost 3000 days of severe symptoms during the reproductive years. Until two decades ago, there were no effective treatments for severe premenstrual syndrome. Even in 2000, almost three-quarters of women in the USA with premenstrual disorders either did not seek help or sought treatment unsuccessfully from at least three clinicians for over 5 years. This review will focus on the epidemiology, diagnosis, treatment outcomes, quality of life and burden of illness for premenstrual disorders.

Rasheed, P., & Al-Sowielem, L. S. (2002) examined the relationship between several possible risk factors, including physical activity, and PMS among college-aged women in Saudi Arabia. They conducted a cross-sectional study with 488 women enrolled in nursing, medical laboratory technology and respiratory therapy educational programs who were between the ages of 17 and 27. They collected information on risk factors through a self-administered questionnaire that asked about level of physical activity (none, daily housework of at least 2 hours, three weekly sessions of walking or aerobic/rhythmic exercises for at least 20 minutes, or participation in sports). Women were also asked to report whether they had experienced any of 14 menstrual symptoms 9 during the last 6 months for 3 or more cycles within one week prior to the onset of menstruation. Scores were assigned based on whether the symptom was described as mild (score=1; daily activities not affected and no medication taken), moderate (score=2; daily activities not affected but medication was taken), or severe (score=3; daily activities were affected despite medication). Each woman was assigned a PMS symptom score from 0 (no symptoms) to 42 (maximum of 14 severe symptoms). The authors found that 96.6% of women in the study reported experiencing at least one menstrual symptom within the last 6 months. Findings from a univariate analysis show that physically active women were more likely to have high PMS symptom scores compared to sedentary women, with 45.3% of physically active women having high ranged scores (10-33) compared to 30.6% of sedentary women ($P<0.01$). It is unclear how the women were divided into physically active and sedentary groups, as there was no description of this categorization, and no further statistical analysis on the difference between the two groups was conducted. Due to the cross-sectional design, temporal bias is an issue for this study, as it is unclear whether women experienced PMS and started or stopped exercising to alleviate symptoms, or exercised regularly and never experienced PMS. The total score for PMS symptoms was also arbitrarily categorized into three levels (0, 1- 9, ≥10) without any explanation for this process. Because information was collected by questionnaire, misclassification of outcome is a limitation of the study as the question on PMS asked about symptoms during the last 6 months, which can be difficult to remember at one point in time.

Reavenall, S., & Blake, H. (2010) assessed whether college students in various stages of change relative to exercise adherence would rate self-efficacy and barriers to exercise differently. A total of 198 participants (49 men, 149 women) ranging in age from 17 to 38 years (M=20.4, SD=2.3) completed surveys measuring exercise behavior, barriers to exercise, and both task and coping self-efficacy. Multivariate tests yielded significant differences in means on task self-efficacy, coping self-efficacy, and perceived barriers to exercise. Implications are that people who do not exercise or exercise sporadically perceive more barriers to exercise than those who exercise regularly. Also, people who exercise regularly rate their confidence higher (i.e., self-efficacy) to overcome obstacles to exercise. Health practitioners may find useful assessments of baseline self-efficacy and perceived barriers before implementing exercise programs.

Ries, A. V., Voorhees, C. C., & Gittelsohn, J. (2010) identified environmental characteristics that act as barriers and facilitators of physical activity among urban African-American adolescents—a population with disproportionately low levels of physical activity. The study used concept mapping, a participatory research methodology, with 50 adolescents in Baltimore, Maryland to identify the strongest environmental barriers and facilitators. Young women identified social environmental factors (e.g., crime, violence, drugs) as their strongest barriers to physical activity. For young men, other barriers (e.g., transportation, finances, and parental influences) were equally as strong as social environmental factors. Social support and neighborhood locations for physical activity were identified as facilitators of physical activity, with social support being the strongest facilitator for male and female participants. This study identifies age and culturally relevant environmental characteristics that should be examined in quantitative studies and considered in future interventions addressing physical activity in a high-risk population.

Ries, A. V., Voorhees, C. C., & Gittelsohn, J. (2010) identified environmental characteristics that act as barriers and facilitators of physical activity among urban African-American adolescents—a population with disproportionately low levels of physical activity. The study used concept mapping, a participatory research methodology, with 50 adolescents in Baltimore, Maryland to identify the strongest environmental barriers and facilitators. Young women identified social environmental factors (e.g., crime, violence, drugs) as their strongest barriers to physical activity. For young men, other barriers (e.g., transportation, finances, and parental influences) were equally as strong as social environmental factors. Social support and neighborhood locations for physical activity were identified as facilitators of physical activity, with social support being the strongest facilitator for male and female participants. This study identifies age and culturally relevant environmental characteristics that should be examined in quantitative studies and considered in future interventions addressing physical activity in a high-risk population.

Rocha Filho, E. A., Lima, J. C., Neto, J. S. P., & Montarroyos, U. (2011) evaluated the effectiveness and safety of polyunsaturated fatty acids for the treatment of the premenstrual syndrome (PMS) using a graded symptom scale and to assess the effect of this treatment on basal plasma levels of prolactin and total cholesterol. A randomized, double-blind, placebo-controlled study was conducted with 120 women with PMS divided into three groups and treated with 1 or 2 grams of the medication or placebo. Symptoms were recorded over a 6-month period using the Prospective Record of the Impact and Severity of Menstruation (PRISM) calendar. Total cholesterol and prolactin levels were measured. Analysis of variance (ANOVA), Pearson's chi-square test, Wilcoxon's nonparametric signed-rank test for paired samples and the Mann-Whitney nonparametric test for independent samples were used in the statistical analysis. The difference between the groups using the medication and the placebo group with respect to the improvement in symptomatology appears to indicate the effectiveness of the drug. Improvement in symptoms was higher when the 2-gram dose was used. This medication was not associated with any changes in prolactin or total cholesterol levels in these women.

Rodio, A., & Fattorini, L. (2014) evaluated the influence of downhill walking on leg strength in young adult. A total of 32 females (age 26 ± 4 years; height 1.64 ± 0.05 m; body mass 57.6 ± 5.6 kg) were divided into four groups and they carried out an exercise intervention consisting of three sessions per week for 6 weeks, each lasting 30 minutes. Groups were defined at several workloads characterised by treadmill inclination (%) and walking speed ($m \cdot s^{-1}$): Level Walking at treadmill inclination 0% and walking speed 1.0; Uphill Walking at +20%, 0.75; Downhill Walking (DW) at -20%, 1.36; and Mixed Walking at +20%, 0.75 and -20%, 1.36 each lasting 15 minutes. Maximum voluntary contraction (MVC) developed by the Quadriceps Femoris and Endurance Time at 60% MVC were evaluated before and after experimental period. At the end of each session, Borg's scale and Visual Analogue Scale (VAS) were adopted in order to evaluate perception of rate exertion and pain. Statistical analysis showed significant only in MVC for DW in both right and left legs. Borg's scale and VAS described light activity free of pain. Present findings showed how an eccentric exercise, short lasting and at a low workload, can be useful in inducing improvements in leg strength.

Romans, S., Clarkson, R., Einstein, G., Petrovic, M., & Stewart, D. (2012) examined one pervasive idea, that the MC is a cause of negative mood, by studying the scientific literature as a whole. We briefly reviewed the history of the idea of premenstrual syndrome and undertook a systematic review of quality studies. We searched PubMed, PsycINFO, and article bibliographies for published studies using non-help-seeking samples with daily mood data collected prospectively for a minimum of 1 complete MC. We critiqued their methodologies and tabulated the key findings. Taken together, these studies failed to provide clear evidence in support of the existence of a specific premenstrual negative mood syndrome in the general population. This puzzlingly widespread belief needs challenging, as it perpetuates negative concepts linking female reproduction with negative emotionality.

Rosenbaum, S., Tiedemann, A., Sherrington, C., Curtis, J., & Ward, P. B. (2014) determined effects of physical activity on depressive symptoms (primary objective), symptoms of schizophrenia, anthropometric measures, aerobic capacity, and quality of life (secondary objectives) in people with mental illness and explore between-study heterogeneity.Randomized controlled trials of adults with a DSM-IV-TR, ICD-10, or clinician-confirmed diagnosis of a mental illness other than dysthymia or eating disorders were selected. Interventions included exercise programs, exercise counseling, lifestyle interventions, tai chi, or physical yoga. Study methodological quality and intervention compliance with American College of Sports Medicine (ACSM) guidelines were also assessed. Physical activity reduced depressive symptoms in people with mental illness. Larger effects were seen in studies of poorer methodological quality. Physical activity reduced symptoms of schizophrenia and improved anthropometric measures, aerobic capacity, and quality of life among people with mental illness.

Salamat, S., Ismail, K. M., & O'Brien, S. (2008) discussed premenstrual syndrome (PMS) is a serious condition that is still poorly understood and accepted. It is a psychological and somatic disorder of unknown aetiology. It is quoted that 95% of women suffer from premenstrual symptoms and 5% of these suffer from PMS. Symptoms of PMS must arise in the luteal phase of a woman's cycle, be relieved by menstruation, and be recurrent and severe enough to have a major impact on normal functioning. There are no tests to diagnose PMS; instead, a careful history must be taken, ruling out differentials. However, the diagnosis should be made following prospective symptom rating using validated tools. Several treatment options have been suggested for PMS, including behavioural therapies, committing to a regular exercise programme, simple dietary alterations, and pharmacological and surgical interventions. However, the most effective treatments tend to fall into one of two categories: suppressing ovulation or correcting a speculated neuroendocrine anomaly.

Schatz, D. B., Hsiao, M. C., & Liu, C. Y. (2012) summarized the literature regarding Premenstrual Dysphoric Disorder (PMDD) in the East Asian countries of Taiwan, China (including Hong Kong and Macau), Japan, and Korea. A Pubmed and Chinese Electronic Periodical Service (CEPS) literature review was conducted using the key words "Premenstrual Dysphoric Disorder" along with "Japan," rea," "Taiwan," "Hong Kong," and "Macau." Using these criteria, 17 articles were found. Three articles were excluded because they did not involve PMDD in the aforementioned countries. In addition to this search, an article found in a review of the research on reproductive mental health disorders in China was utilized. That review contained one article regarding PMDD with an English language abstract, which was utilized in this current article. The rates of PMDD appear to be lower in East Asia, though reasons for this result are discussed. Few studies have been conducted examining the efficacy of psychotropic medications commonly used to treat PMDD in this population.

Shah, N. R., Jones, J. B., Aperi, J., Shemtov, R., Karne, A., & Borenstein, J. (2008) severed premenstrual syndrome (PMS) affects between 3% to 5% of women of reproductive age. Severe PMS is classified under the Diagnostic and Statistical Manual of Mental Disorders as premenstrual dysphoric disorder (PMDD). Selective serotonin reuptake inhibitors (SSRIs) are increasingly used as front-line therapy for PMS. A systematic review was undertaken on the efficacy of SSRIs in the management of severe PMS, or PMDD, to assess the evidence for this treatment option. The objective of this review was to evaluate the effectiveness of SSRIs in reducing premenstrual syndrome symptoms in women diagnosed with severe premenstrual syndrome. All trials were considered in which women with a prospective diagnosis of PMS, PMDD or late luteal phase dysphoric disorder (LPDD) were randomised to receive SSRIs or placebo for the treatment of premenstrual syndrome in a blinded trial. Forty randomised controlled trials were identified which reported the use of SSRIs in the management of PMS. Fifty-six trials were excluded. The review authors extracted the data independently and estimated standardised mean differences for continuous outcomes.The evidence supports the use of selective serotonin reuptake inhibitors in the management of severe premenstrual syndrome.

Shah, N. R., Jones, J. B., Aperi, J., Shemtov, R., Karne, A., & Borenstein, J. (2008) reviewed evidence of the treatment benefits of selective serotonin reuptake inhibitors (SSRIs) for symptoms related to severe premenstrual syndrome (PMS) and premenstrual dysphoric disorder.Studies included in the review were double-blind, randomized, controlled trials comparing an SSRI with placebo that reported a change in a validated score of premenstrual symptomatology. Studies had to report follow-up for any duration longer than one menstrual cycle among premenopausal women who met clinical diagnostic criteria for PMS or premenstrual dysphoric disorder. From 2,132 citations identified, we pooled results from 29 studies (in 19 citations) using random-effects meta-analyses and present results as odds ratios (ORs).Selective serotonin reuptake inhibitors were found to be effective in treating premenstrual symptoms, with continuous dosing regimens favored for effectiveness.

Sharifi, F., Simbar, M., Mojab, F., & Majd, H. A. (2014) compared the effects of Chamomile Extract and Mefenamic acid (MA) on the intensity of Premenstrual syndrome (PMS) symptoms.This study was a clinical randomized double blind trial, carried out with 90 students living in the dorms of Iran. The participants filled the daily forms about the intensity of PMS for two consecutive months. Once the definitive diagnosis of PMS was made, the participants were divided into two groups, each receiving either Chamomile capsule 100 mg or MA 250 mg three times a day. Intensity reduction of emotional symptoms was significantly higher among Chamomile Extract-users (30.1 ± 26.6 and 33.4 ± 25.3 percent) than that among MA-users (11.6 ± 25.7 and 10.7 ± 26.8 percent) after two cycles intervention ($p < 0.001$). Intensity reduction of physical symptoms was not significantly different ($p > 0.05$) among groups. Consumption of Chamomile seems to be more effective than MA in relieving the intensity of PMS associated symptomatic psychological pains.

Sharifirad, G., Charkazi, A., Tashi, M., Shahnazi, H., & Bahador, E. (2011) investigated of physical activity distribution bases on stages of change among Isfahan University of Medical Sciences central part of Iran. Exercise behavior stages of change construct questionnaire were collected from 504 participants by using a convenience sample in May 2010. The majority of the university students are suffering physical inactivity that need to design more effective physical activity promotion programs in the community and universities level.

Shulman, L. P. (2010) discussed the vast majority of menstruating women experience uncomfortable symptoms during the premenstrual phase of their menstrual cycles. Although many women do not require specific treatment of their symptoms, approximately 20% to 50% report moderate to severe premenstrual symptoms and about 5% meet the diagnostic criteria for premenstrual dysphoric disorder, the most severe manifestation of premenstrual symptoms. While the etiology of premenstrual symptoms remains unclear, several theories have implicated sex steroids and neurotransmitters in the development and manifestation of symptoms. Further complicating the delineation of etiology is that premenstrual symptoms can be somatic, psychological, or behavioral, as well as a combination of all three. Developing successful interventions for premenstrual symptoms has thus been challenging, with interventions focused on a particular aspect of premenstrual symptomatology. Treatments for premenstrual symptoms include lifestyle changes, cognitive behavioral therapies, and pharmacologic agents including ovulation suppression regimens, antidepressant medications, and anxiolytics.

Shuval, K., Weissblueth, E., Amira, A., Brezis, M., Faridi, Z., Ali, A., & Katz, D. L. (2008) designed to increase physical activity in this population have been scarce. To improve our understanding of the cultural, religious, and environmental barriers and enablers to physical activity, we conducted a qualitative study among Arab Israeli college students in Israel. A total of 45 students participated in 8 focus groups. Purposeful sampling was used to capture the diverse characteristics of participants. Two researchers analyzed the data independently guided by grounded theory. Peer-debriefing sessions were held to group preliminary categories into larger themes. Generally, consensus between researchers was high, and minor differences were resolved. Increasing Arab Israelis' access to safe and culturally appropriate exercise facilities should become a priority. Thus, policy changes in allocating appropriate funds to promote physical activity must be considered, along with using multiple health promotion strategies.

Sigrist, L. D., Anderson, J. E., & Auld, G. W. (2005) identified the need for more appropriate nutrition and fitness education for military personnel. The purpose of this study was to assess senior military officers' concerns on various health topics, educational preferences for nutrition and health topics, eating habits, and barriers and motivators for eating healthfully and exercising regularly. The survey was completed by 52 resident students at the U.S. Army War College. Fitness, weight, and blood cholesterol were top health concerns, and respondents wanted to know more about eating healthfully on the run. The primary barrier to eating healthfully and exercising regularly was lack of time, whereas health and appearance were top motivators. Health interventions for this population should include their topics of concern and should address perceived barriers and motivators.

Sleep, I. (2007) investigated sleep quality and sleep composition using conventional and quantitative electroencephalographic analyses in women with severe PMS, as compared with that of controls. Women (aged 18–40 years) were screened to ensure that their PMS symptoms were severe and that they had ovulatory menstrual cycles. Nine women with PMS or PMDD and 12 asymptomatic control subjects then had laboratory-based polysomnographic recordings at 2 phases of the menstrual cycle: follicular phase and late luteal (premenstrual) phase. Perceived poor quality sleep is a characteristic of severe PMS, but sleep composition based on polysomnographic measures and quantitative electroencephalographic analysis does not differ in association with premenstrual symptom expression in the late luteal phase.

Smith, L. L., Larkey, L. K., Roe, D. J., Bucho-Gonzalez, J. A., Saboda, K., & Ainsworth, B. E. (2014) discussed adherence to recommendations for physical activity (PA) among adults in the United States is reported as poor, particularly among low-income Latinos. In particular, Latina women are at increased risk for inactivity and chronic illness, but national health survey reports aggregate data across demographic strata, limiting descriptive information that could guide targeted PA promotion. The purpose of the study was to determine PA patterns among a low-income group of Latina women in the southwestern United States.

Sohrabi, N., Kashanian, M., Ghafoori, S. S., & Malakouti, S. K. (2013) evaluated the effect of omega-3 fatty acids on the treatment of PMS.A randomized double blind controlled trial was performed on 184 eligible women. The eligible women were randomly assigned into two groups. The number of women who have finalized the study with us was 124.In the case group (omega-3 group = group A, $n = 70$), omega-3 in an amount of 2 g was prescribed for a one per day basis on a single dosage (two 1 g pearls), and in the control group (placebo group = group B, $n = 69$) 2 placebo soft gel, which were completely similar to omega-3 soft gels, were prescribed. The severity and duration of each of the symptoms were compared in both groups 1.5 and 3 months after the beginning of treatment. It appears that omega-3 fatty acids may reduce the psychiatric symptoms of PMS including depression, nervousness, anxiety, and lack of concentration and may also reduce the somatic symptoms of PMS including bloating, headache and breast tenderness. These effects increased by longer duration of treatment.

Stamatelopoulos, K. S., Georgiopoulos, G., Papaioannou, T., Lambrinoudaki, I., Kouzoupis, A., Vlachopoulos, C., ... & Sfikakis, P. P. (2012) tested the hypotheses that monthly fluctuations in markers of arterial stiffness and blood pressure hemodynamics differ between women with and without premenstrual syndrome. We also assessed hypertension prevalence and arterial stiffening in postmenopausal women with or without history of premenstrual symptoms. Twenty one pre-menopausal women with premenstrual syndrome and 15 women without were prospectively examined in three distinct phases of their menstrual cycle (menses, late follicular and luteal phase). Pulse-wave velocity and analysis were used to assess arterial stiffness and wave reflection indices, respectively. Endothelial function was evaluated by flow-mediated vasodilation. In a cross-sectional substudy, 156 postmenopausal women were assessed for possible associations between retrospectively reported PMS symptoms and hypertension. results imply that PMS may affect arterial stiffness and BP monthly variability. Whether PMS is associated with new onset hypertension later in life needs further evaluation.

Steiner, M., Peer, M., Macdougall, M., & Haskett, R. (2011) proposed to bring them in line with the DSM-IV criteria for Premenstrual Dysphoric Disorder (PMDD) by updating the PMTS-O and replacing the PMTS-SR with a Multiple Visual Analogue Scale (PMTS-VAS).A convenience sample of 23 Caucasian, English-speaking women in their reproductive years with regular menstrual cycles was recruited. Participants were administered the revised PMTS-O (PMTS-OR) by a trained clinician and then instructed to complete the PMTS-SR and the new PMTS-VAS, both of which were timed. The participants were also asked which of the instruments they preferred.By making minor modifications to the PMTS-O we have ensured that all criteria for the DSM-IV definition of PMDD are now represented in the PMTS-OR. The new PMTS-VAS mirrors the PMTS-OR but now also captures the severity of self rated symptoms. These scales are simple to complete for both clinicians and clients, and are reliable, valid and sensitive to change.

Stice, E., Rohde, P., Shaw, H., & Marti, C. N. (2012) evaluated a selective prevention program targeting both eating disorder symptoms and unhealthy weight gain in young women. Female college students at high-risk for these outcomes by virtue of body image concerns (N = 398; M age = 18.4 years, SD = 0.6) were randomized to the Healthy Weight group-based 4-hr prevention program, which promotes gradual lasting healthy improvements to dietary intake and physical activity, or an educational brochure control condition. Results indicate that this intervention reduced both eating disorder symptoms and unhealthy weight gain, but suggest it should be improved to produce stronger and more persistent effects, and that it may be useful to target young women with both body image and eating disturbances.

Swartz, A. M., Rote, A. E., Cho, Y. I., Welch, W. A., & Strath, S. J. (2014) determined the responsiveness of two motion sensors to detect change in sedentary behaviour (SB) and physical activity (PA) during an occupational intervention to reduce sitting time.SB and PA were assessed using a hip-worn Actigraph GTX3 (AG) and a thigh-worn activPAL (AP) during three consecutive workdays throughout baseline and intervention periods. Mean scores at baseline and intervention were estimated by hierarchical linear models (HLM) with robust SEs, adjusting for random variance of average scores between participants. Change scores (mean baseline minus mean intervention) were calculated for each device. Response to change was assessed for each device using the standardised response mean. Responsiveness to change in SB and PA was similar and comparable for the AP and AG, indicating agreement across both measurement devices.

Taguchi, R., Matsubara, S., Yoshimoto, S., Imai, K., Ohkuchi, A., & Kitakoji, H. (2009). characterized mainly by psychological symptoms confined to the premenstrual period, which reduce not only patients' quality of life, but also their working activities. Although various therapies have been employed for PMDD, some patients do not respond to them. We recently employed acupuncture treatment for a patient in PMDD. Symptoms ameliorated during the acupuncture (+) period, but deteriorated during the acupuncture (-) period. This review describes the clinical course of this case. The diagnosis and treatment of PMDD are briefly summarized and previous acupuncture treatment for PMS are reviewed. The difficulties in evaluating the effectiveness of acupuncture for PMS/PMDD are addressed. It is suggested that acupuncture may be a treatment option for PMDD.

Teixeira, A. L. D. S., Oliveira, É. C. M., & Dias, M. R. C. (2013) determined the relationship between the level of physical activity and the incidence of premenstrual syndrome. A cross-sectional design was conducted on 71 apparently healthy university students (24.4±4.8 yrs; 61.5±8.7 kg; 1.63±0.06 m). The level of physical activity was determined with a questionnaire and the presence of premenstrual syndrome was verified based on daily symptoms self-reported in a diary during two consecutive menstrual cycles. 17 premenstrual symptoms are considered in the diary, which should be scored on a 5-point scale (0-4) according to their occurrence, so that a score can be calculated in each cycle. The occurrence of premenstrual syndrome was considered if three or more symptoms were reported up to six days before menstruation (premenstrual period) and were absent up to six days after menstruation (postmenstrual period).There is a negative relationship between the level of physical activity and the incidence of premenstrual syndrome, with women with a positive diagnosis of premenstrual syndrome having a lower level of physical activity than healthy women.

Terry- McElrath, Y. M., & O'malley, P. M. (2011) examined the extent to which the trajectory of participation in sports, athletics or exercising (PSAE) covaried with substance use in early adulthood controlling for team sports participation using parallel process latent growth curve modeling. Analysis of data collected from a series of panel studies using a cohort-sequential design. Specifically, the analyses used longitudinal data from 11 741 individuals from the graduating classes of 1986-2001, first surveyed as seniors in American high schools. Up to four additional follow-up surveys were administered to age 26 years. Data were collected using in-school and mailed self-administered questionnaires. Level of PSAE, past-30-day alcohol, cigarette and marijuana use frequency and any past-30-day use of illicit drugs other than marijuana (IOTM) were the main processes of interest. Self-reported race/ethnicity, college status at age 19/20 years, parental education, gender and team sports participation during high school were included as covariates. Increased participation in sports, athletics or exercising (PSAE) is related to significantly lower substance use frequency at modal age 18 and through significantly and negatively correlated growth trajectories through early adulthood. Encouraging PSAE among adolescents and early adults may relate to lower substance use levels throughout early adulthood.

Thompson Coon, J., Boddy, K., Stein, K., Whear, R., Barton, J., & Depledge, M. H. (2011) compared the effects on mental and physical wellbeing, health related quality of life and long-term adherence to physical activity, of participation in physical activity in natural environments compared with physical activity indoors. We conducted a systematic review using the following data sources: Medline, Embase, Psychinfo, GreenFILE, SportDISCUS, The Cochrane Library, Science Citation Index Expanded, Social Sciences Citation Index, Arts and Humanities Citation Index, Conference Proceedings Citation Index--Science and BIOSIS from inception to June 2010. Internet searches of relevant Web sites, hand searches of relevant journals, and the reference lists of included papers and other review papers identified in the search were also searched for relevant information. Controlled trials (randomized and nonrandomized) were included. To be eligible trials had to compare the effects of outdoor exercise initiatives with those conducted indoors and report on at least one physical or mental wellbeing outcome in adults or children. Screening of articles for inclusion, data extraction, and quality appraisal were performed by one reviewer and checked by a second with discrepancies resolved by discussion with a third if necessary. Due to the heterogeneity of identified studies a narrative synthesis was performed. Eleven trials (833 adults) were included. Most participants (6 trials; 523 adults) were young students. Study entry criteria and methods were sparsely reported. All interventions consisted of a single episode of walking or running indoors with the same activity at a similar level conducted outdoors on a separate occasion. A total of 13 different outcome measures were used to evaluate the effects of exercise on mental wellbeing, and 4 outcome measures were used to assess attitude to exercise. Most trials (n = 9) showed some improvement in mental wellbeing on one or other of the outcome measures. Compared with exercising indoors, exercising in natural environments was associated with greater feelings of revitalization and positive engagement, decreases in tension, confusion, anger, and depression, and increased energy. However, the results suggested that feelings of calmness may be decreased following outdoor exercise. Participants reported greater enjoyment and satisfaction with outdoor activity and declared a greater intent to repeat the activity at a later date. None of the identified studies measured the effects of physical activity on physical wellbeing or the effect of natural environments on exercise adherence. The hypothesis that there are added beneficial effects to be gained from performing physical activity outdoors in natural environments is very appealing and has generated considerable interest. This review has shown some promising effects on self-reported mental wellbeing immediately following exercise in nature which are not seen following the same exercise indoors. However, the interpretation and extrapolation of these findings is hampered by the poor methodological quality of the available evidence and the heterogeneity of outcome measures employed. The review demonstrates the paucity of high quality evidence on which to base recommendations and reveals an undoubted need for further research in this area. Large, well designed, longer term trials in populations who might benefit most from the potential advantages of outdoor exercise are needed to fully elucidate the effects on mental and physical wellbeing. The influence of these effects on the sustainability of physical activity initiatives also awaits investigation.

Tomporowski, P. D., Lambourne, K., & Okumura, M. S. (2011) provided a historical overview of physical activity interventions designed by American educators and an evaluation of research that has assessed the effects of exercise on children's mental function.Historical descriptions of the emergence of American physical education doctrine throughout the 20th century were evaluated. Prior reviews of studies that assessed the effects of single acute bouts of exercise and the effects of chronic exercise training on children's mental function were examined and the results of recent studies were summarized. Physical activity interventions for children should be designed to meet multiple objectives; e.g., optimize physical fitness, promote health-related behaviors that offset obesity, and facilitate mental development.

Usman, S. A. B., Indusekhar, R., & O'Brien, S. (2008) discussed remenstrual syndrome (PMS) is a psychological and somatic disorder of unknown aetiology. The symptoms of PMS regularly occur during the luteal phase of the menstrual cycle and resolve by the end of menstruation. The severe and predominantly psychological form of PMS is called 'premenstrual dysphoric disorder'. PMS results from ovulation and appears to be caused by the progesterone produced following ovulation in women who have enhanced progesterone sensitivity. This enhanced sensitivity may be due to neurotransmitter dysfunction. Treatment is aimed at suppressing ovulation or reducing progesterone sensitivity. This chapter will describe the role of hormones and hormonal treatments in PMS.

Van der Leij, F., Schultz, W. C., van de Wiel, H., & van Leeuwen, J. H. (2009) determined the premenstrual syndrome (PMS) is characterized by the occurrence of physical and psychological symptoms during the luteal phase of almost every menstrual cycle. These symptoms disappear at the beginning of menstruation, and a symptom-free period of at least a week ensues. Premenstrual dysphoric disorder (PMDD) is a variation of PMS, with predominantly psychological symptoms. The aetiology of PMD and PMDD is not known. A possible explanation however is an abnormal, stronger reaction to physiologically normal hormonal fluctuations. Diagnosing PMS and PMDD requires prospective daily monitoring of symptoms over at least two menstrual cycles. No effective medication for the treatment of PMS has been registered in the Netherlands. In randomized placebo-controlled trials selective serotonin reuptake inhibitors and oral contraceptives containing drosperinone have been found to have a positive effect on the physical and psychological symptoms of PMS and PMDD.

Vander Ploeg, K. A., Maximova, K., McGavock, J., Davis, W., & Veugelers, P. (2014) compared the two-year change in physical activity among 10–11 year-old children attending schools with and without health promotion programs by activity level, body weight status, and socioeconomic backgrounds to assess whether health promotion programs reduce or exacerbate health inequalities. This was a quasi-experimental trial of a Comprehensive School Health (CSH) program implemented in schools located in socioeconomically disadvantaged neighbourhoods in Edmonton, Alberta, Canada. In the spring of 2009 and 2011, pedometer (7 full days) and demographic data were collected from cross-sectional samples of grade five children from 10 intervention and 20 comparison schools. Socioeconomic status was determined from parent self-report. Low-active, active, and high-active children were defined according to step-count tertiles. Multilevel linear regression methods adjusted for potential confounders were used to assess the relative inequity in physical activity and were compared between groups and over-time. In 2009, a greater proportion of students in the intervention schools were overweight (38% vs. 31% $p = 0.03$) and were less active (10,827 vs. 12,265 steps/day $p < 0.001$). Two years later, the relative difference in step-counts between intervention and comparison schools reduced from −15.5% to 0% among low-active students, from −13.4% to 0% among active students, and from −15.1% to −2.7% among high-active students. The relative difference between intervention and comparison schools reduced from −11.1% to −1.6% among normal weight students, from −16.8% to −1.4% among overweight students, and was balanced across socioeconomic subgroups. These findings demonstrate that CSH programs implemented in socioeconomically disadvantaged neighbourhoods reduced inequalities in physical activity. Investments in school-based health promotion are a viable, promising, and important approach to improve physical activity and prevent childhood obesity, and may also reduce inequalities in health.

Vaughn, L. M., Jacquez, F., Zhao, J., & Lang, M. (2011) analyzed the School-based health programs have a unique and powerful potential to meet the health needs of children because students spend more time in schools than in any other environment away from home. We conducted a participatory needs assessment called a Group Level Assessment (GLA) in collaboration with an economically disadvantaged and ethnically diverse school to facilitate the students' identification of and subsequent action toward important health needs. A total of 68 students in Kindergarten through eighth grade participated in the GLA. Four major themes emerged: the desire for more sports and after-school activities, better school lunches, enjoyment of friends and families, and overall happiness. Other health issues identified by the students included limited health/medical care, stress resulting from schoolwork and grades, positive self-image, and the desire for more art opportunities. The salient themes identified by students are consistent with many factors identified in the academic literature as important in child socioemotional functioning.

Vichnin, M., Freeman, E. W., Lin, H., Hillman, J., & Bui, S. (2006) Identified the presence of patients with premenstrual syndrome (PMS) in an adolescent gynecology practice, and evaluate the reported severity, impairment and timing of the symptoms in the menstrual cycle. Adolescents ages 13–18 years completed a symptom questionnaire, functional impairment ratings, and a brief medical history questionnaire during an office visit. Teens who responded that they had PMS and reported a premenstrual symptom score at least 50% greater than the postmenstrual score and rated moderate to severe impairment in one or more domains comprised the "PMS" group. Teens who responded that they had PMS but did not meet the symptom and impairment criteria were termed "PMS not supported." Teens who responded that they did not have PMS and did not meet the PMS symptom and impairment criteria were termed "No PMS."The reports of premenstrual symptoms, their severity, timing and impairment suggest that PMS is common in adolescents. Further study is warranted to confirm these results with prospective assessment of PMS and to evaluate treatments for adolescents who have clinically significant PMS.

Vickers, K. S., Patten, C. A., Bronars, C., Lane, K., Stevens, S. R., Croghan, I. T., & Clark, M. M. (2004) investigated the relation of potentially modifiable factors (physical activity level, weight concern, and depressive symptoms) to binge drinking while controlling for the effects of previously established correlates of binge drinking (tobacco and marijuana use, GPA, and perception of peer alcohol use). Four-hundred-twelve college women completed an in-class survey. Multivariate analyses revealed that tobacco and marijuana use, GPA, and physical activity were significantly associated with binge drinking, whereas tobacco use and perception of peers' alcohol use were associated with more frequent binge drinking. The findings suggested that the variables associated with any binge drinking and frequency of binge drinking may differ and that binge drinking can be associated with positive health behaviors (ie, greater physical activity) as well as risky health behaviors (eg, tobacco use).

Vigod, S. N., Frey, B. N., Soares, C. N., & Steiner, M. (2010) examined as many as 7% of women experience significant social or occupational dysfunction as a result of severe premenstrual mood disturbance. Biological, psychological, and sociocultural factors are implicated in the cause of premenstrual dysphoric disorder, but the interaction between these factors remains to be elucidated. Mental health practitioners can aid women by providing diagnostic clarity and by initiating an integrated step-wise management approach.

Wang, Y. J., Hsu, C. C., Yeh, M. L., & Lin, J. G. (2013) studied used objective, physiological measurements to evaluate the effects of auricular acupressure on menstrual pain and menstrual distress in young college students with primary dysmenorrhea across two menstrual cycles. The aim was to determine if significant differences could be detected between the intervention and follow-up phases after controlling life stress. A one-group experimental research design was used, and repeated measurements and follow ups were done. Thirty-two women completed questionnaires and physiological parameters were measured. Auricular acupressure effectively increases parasympathetic activity to maintain autonomic function homeostasis in young women with primary dysmenorrhea and may have a value in alleviating menstrual pain and menstrual distress in a high-stress life. Future studies should consider stress, stimulus dose of auricular acupressure, severity of menstrual pain, and a longitudinal research design.

Washington, W. D., Banna, K. M., & Gibson, A. L. (2014) analyzed an estimated 30% of Americans meet the criteria for obesity. Effective, low-cost interventions to increase physical activity are needed to prevent and treat obesity. In this study, 11 healthy adults wore Fitbit accelerometers for 3 weeks. During the initial baseline, subjects earned prize draws for wearing the Fitbit. During intervention, percentile schedules were used to calculate individual prize-draw criteria. The final week was a return to baseline. Four subjects increased step counts as a result of the intervention. A bout analysis of interresponse times revealed that subjects increased overall step counts by increasing daily minutes active and within-bout response rates and decreasing pauses between bouts of activity. Strategies to improve effectiveness are suggested, such as modification of reinforcement probability and amount and identification of the function of periods of inactivity.

Weisz, G., & Knaapen, L. (2009) determined premenstrual syndrome (PMS) and its derivative Premenstrual Dysphoric Disorder (PMDD) are controversial medical diagnoses. On one hand they are ubiquitous in English-language cultures; on the other they are for many emblematic of unnecessary medicalization of natural physiological processes. In this paper, we use data produced by IMS, a health care information and research firm, to analyze office-based medical practice related to PMS/PMDD in five countries. We come to several Relatively few doctors in any country diagnose women as suffering from PMS/PMDD, despite significant national variations in frequency of diagnosis; Women diagnosed with this condition are usually prescribed a medication no matter what kind of specialist they see; and. In North America and the UK, practitioners generally follow USA practice guidelines which favour use of anti-depressive drugs like SSRIs but this is not the case in France and Germany which exhibit unique prescription patterns. In France hormonal treatment and analgesics dominate; in Germany the plant extract Vitex agnus-castus, considered an alternative therapy in much of the English-speaking world, is most common. We go on to discuss the relevance of these conclusions to discussions of variations in medical practices, to the existing PMS literature that claims high rates of prevalence for this condition, and to recent studies of "demedicalization" in certain domains.

Whelan, A. M., Jurgens, T. M., & Naylor, H. (2009) identified herbs, vitamins and minerals advocated for the treatment of PMS and/or PMDD and to systematically review evidence from randomized controlled trials (RCTs) to determine their efficacy in reducing severity of PMS/PMDD symptoms. Searches were conducted from inception to April 2008 in Clinical Evidence, The Cochrane Library, Embase, IBID, IPA, Mayoclinic, Medscape, MEDLINE Plus, Natural Medicines Comprehensive Database and the Internet to identify RCTs of herbs, vitamins or minerals advocated for PMS. Bibliographies of articles were also examined. Included studies were published in English or

French. Studies were excluded if patient satisfaction was the sole outcome measure or if the comparator was not placebo or recognized therapy. Only calcium had good quality evidence to support its use in PMS. Further research is needed, using RCTs of adequate length, sufficient sample size, well-characterized products and measuring the effect on severity of individual PMS symptoms.

Yan, Z., & Cardinal, B. J. (2013) identified as one of the least physically active groups in the United States. In an effort to better understand this situation, this study's purpose was to examine CFIS in American higher education in terms of the meaning they assigned to physical activity and facilitators and barriers they experienced with regard to physical activity participation. Twenty CFIS from a university in the Northwest region of the United States were recruited and interviewed. All of the interviews were conducted in Mandarin Chinese and translated and transcribed into English. The 1-on-1 semi structured interviews lasted between 45 min and 60 min each. Data were analyzed by NVivo8. Understanding the physical activity experiences of CFIS is an important step in the process of promoting their long-term health and well-being. It may behoove universities with growing Chinese international student populations to add more cross-cultural content into their curriculums and fitness programs, increase awareness of cultural differences on campus, and increase accessibility to information in an effort to remove physical activity participation barriers experienced by CFIS.

Yang, C. C., Brown, B. B., & Braun, M. T. (2014) obtained the 'when, where and why' of healthy bouts of moderate-to-vigorous physical activity (MVPA) provides insights into natural PA. In Salt Lake City, Utah, adults wore accelerometer and Global Positioning System (GPS) loggers for a week in a cross-sectional study to establish baseline travel and activity patterns near a planned Complete Street intervention involving a new rail line, new sidewalks and a bike path. These techniques provide a novel way to understand participant memories of the context and subjective assessments associated with healthy bouts of PA. Prompts with time-stamped maps that illustrate places of MVPA offer an effective method to improve understanding of activity and its supportive sociophysical contexts.

Zaafrane, F., Faleh, R., Melki, W., Sakouhi, M., & Gaha, L. (2007) identified premenstrual syndrome (PMS) is a recurrent disorder that occurs in the luteal phase of the menstrual cycle. It is characterized by intense physical, psychological, and behavioral changes that interrupt interpersonal relationships and disrupt the lives of affected women. Premenstrual syndrome is believed to affect 75% of women of childbearing age. Because no tests can confirm PMS, the diagnosis should be made on the basis of a patient-completed daily symptom calendar and the exclusion of other medical disorders. PMS symptoms occur during the luteal phase of the menstrual cycle and remit with the onset of menstruation or shortly afterward. The aetiology of PMS is still unknown uncertain, but are likely associated with aberrant responses to normal hormonal fluctuations during the menstrual cycle. A wide range of therapeutic interventions has been tested in the treatment of premenstrual symptoms. Most non-pharmacological interventions that have been proven efficacious require a series of interventions. If non-medical approaches are ineffective, drug therapy may be appropriate. Several pharmaceutics agents have been shown to relieve symptoms. Calcium carbonate and selective serotonin reuptake inhibitors have demonstrated excellent efficacy.

Zabinski, M. F., Calfas, K. J., Gehrman, C. A., Wilfley, D. E., & Sallis, J. F. (2001) teached university seniors behavioral skills necessary for increasing and/or maintaining physical activity habits in preparation for the transition to working adult roles after graduation. This study examines the secondary effects of this intervention on body image concerns among college-aged men and women. Three hundred thirty-eight undergraduates (54%female, Mage = 24years, SD = 1.95; MBody Mass Index = 24.26, SD = 4.0) were studied. The sample was 61/% Anglo, 16% Latino, 16% Asian/Pacific Islander, 4% African American, and 3% Native American/Other Body image concerns were assessed at pre- and posttreatment using 2 subscales of the Eating Disorder Inventory: Drive for Thinness and Body Dissatisfaction. Because the latter concentrates on body parts typically associated with female concerns (e.g., thighs, hips, buttocks), a parallel scale was developed to target body parts that may be of more concern to men (e.g., legs, shoulders, arms, stomach). Results indicated that compared to the control group, women in the intervention showed a significant increase in drive for thinness without any changes in body dissatisfaction. For men, there were no significant changes in drive for thinness or body dissatisfaction. These results suggest that physical activity interventions may have some negative consequences of increasing concerns about thinness in women. This negative effect occurred despite intervention content designed to prevent concern over eating, dieting, and the importance of weight. Health promotion studies should include assessments of potential negative side effects.

Zhu, X., & Lee, C. (2009) identified multi-level correlates of walking to/from school and relevant policy implications. We surveyed parents/guardians of 2,695 students from 19 elementary schools in Austin, Texas, which featured diverse sociodemographic and environmental characteristics. Among the personal and social factors, negative correlates were parents' education, car ownership, personal barriers, and school bus availability; positive correlates were parents' and children's positive attitude and regular walking behavior, and supportive peer influences. Of physical environmental factors, the strongest negative correlates were distance and safety concerns, followed by the presence of highways/freeways, convenience stores, office buildings, and bus stops en route. Our findings suggest that society should give high priority to lower socioeconomic status populations and to multi-agency policy interventions that facilitate environmental changes, safety improvements, and educational programs targeting both parents and children.

Zhu, X., Arch, B., & Lee, C. (2008) examined the prevalence and correlates of walking to or from school in eight elementary schools in Austin,

Texas, which have high percentages of low-income, Hispanic students. A survey of 1,281 parents was conducted, including questions about personal, social, and environmental factors that may influence their decisions on the children's school transportation. Binary logistic regressions were used to estimate the odds of choosing walking as the children's typical school travel mode. The results showed that walking was a typical mode for 28 and 34% of trips to and from school, respectively, and mostly accompanied by an adult. Parents' education level, family's car ownership, children's and parents' personal barriers, and having the school bus service reduced the likelihood of walking, while positive peer influences encouraged walking. Among the physical environmental factors, living close to school was the strongest positive predictor; safety concerns and the presence of highway or freeway en route were negative correlates. We concluded that the location of school is a key, as it determines the travel distance and the presence of highway or freeway en route. In addition to environmental improvements, educational and other assistance programs are needed for both parents and children to overcome their personal barriers and safety concerns. Health and disparity issues require further attention, as many underprivileged children have no other means of school transportation but walking in unsafe and poor environments.

2.3 Summary of Literature

The review of literature helped the investigator to spot out relevant topics and variables. Further the literature helped the investigator to frame the suitable hypothesis leading to the problems. The latest literature also helped the investigator to support her findings with regard to the problem. Further the literature collected in the study will also help the research scholar understanding in the similar areas.

The reviews were presented under the one sections such as studies on physical activity barrier, premenstrual syndrome and health status with chronological and alphabetical order. All the research studies were presented in the section proves that there is a significant improvement on dependent variables due to physical activity intervention.

The research studies reviewed are from many journals available in the websites such as Pub Med, Science Direct Journals, ERIC websites etc., employ the dependent variables such as physical activity barrier, premenstrual syndrome and health status among college women student.

The review of literature helped the researcher from the methodological point of view too. It was learnt that most of the research studies cited in this chapter on the varied physical activity intervention would effectively improve the various parameter.

Chapter III

METHODOLOGY

3.1 INTRODUCTION

This chapter describes the procedures followed in the selection of subjects, selection of variables, selection of tools, selection of test, reliability of the data, reliability and validity of the questionnaire, collection of the data, the experimental design and the method adopted for statistical treatment of the data have been explained.

The purpose of the study was to analyse the level of physical activity, premenstrual syndrome, health status and physical activity intervention among the college women students. The manner of handing the various details of the experiment is extremely important to the success of research. Research methodology involves the systematic procedures by which the researcher starts from the initial identification of the problem to its conclusion.

3.2 SELECTION OF SUBJECTS

To conduct this study, initially 1000 female were selected from the affiliated colleges of Manonmanian Sundaranar University, Tirunelveli at random. Five women Arts and Science Colleges were selected at random from Tirunelveli districts and these Colleges were Rani Anna College –Tirunelveli, Sarah Tucker College – Tirunelveli, Sri Sarada College – Tirunelveli, Rose Mary College - Tirunelveli and Sri Parasakthi College - Coutralam. From each of this college 200 students were selected randomly and their age ranged from 18 to 21 years. Further 30 students were selected randomly from Rani Anna College, Tirunelveli for experiment. The selected subjects were randomly assigned into two groups of (n=15) each, such as experimental and control groups. Group I (n=15) underwent physical activity intervention programme for a duration of 12 weeks and the number of sessions per week was confined to three alternative days, in addition to the regular schedule of the curriculum. Group II (n=15) acted as control, were was asked to refrain from any special training except their leisure time pursuit as college students.

3.3 SELECTION OF VARIABLES

As per the availability literature of and the personal knowledge of the investigator following variables were be chosen and for the present study.

Dependent variables

1. Level of Physical activity
2. Premenstrual syndrome
3. Health status

Independent variables

1. Physical activity intervention

3.4 Selection of Tools

As per the available literature the selected variables were tested by using the following standardized tests and they were presented in Table 3.1.

Table 3.1
SELECTION OF TESTS

Sl. No	Variables	Tests	Units of Measurement
1	Level of Physical activity	Questionnaire	In points
2	Premenstrual syndrome	Questionnaire	In points
3	Health status	Questionnaire	In points

3.5 ORIENTATION TO THE SUBJECTS

The investigator explained the purpose of the study to the subjects and their part in the study. For the collection of data in the investigator explained the procedure of testing on selected variable and gave instructions about the procedures to be adopted by them for measuring. All the test materials were printed in a booklet form. Instructions were given on the top of each questionnaire as to how to answer the questions. The investigator distributed the questionnaires to the subjects and assured them that the information gathered from them would be kept confidential. After distributing the questionnaires the investigator explained to all the subjects the procedure to fill up the questionnaire. They were asked to read each statement carefully and then select any one of the responses that fit to them best. The investigator clarified the doubts of the subjects whenever they asked.

Further Four sessions were spent to familiarize the subjects with the technique involved to execute the physical activity intervention. It helped them to perform the given training perfectly and avoids injuries, further; the control group was specially oriented, advised and controlled to avoid the special practice of any of the specific training programme till the end of the experimental period. The participants of all the groups were sufficiently motivated to perform their maximal level during training and testing periods.

3.6 COMPETENCY OF THE TESTER

All the measurement in this study was taken by the investigator with assistance of director of physical education from Rani Anna Govt College for women, Tirunelveli District. To ensure that the investigator and his assistance were well versed with the techniques of conducting tests, they had a number of practice sessions in the correct testing procedure. The tester's reliability was established by test and re-test method.

3.7 RELIABILITY OF THE DATA

The reliability of the data was established by test and retest method was followed in order to establish the reliability of the data by using ten subjects at random. All the dependent variables selected in the present study were tested twice by the same personnel under similar conditions. The intraclass coefficient of correlation was used to find out the reliability of the data and the results are presented in Table 3.2

Table 3.2
Intraclass Co-Efficient of Correlation on Selected Variables

Sl. No	Variables	R -Value
1	Physical activity barrier	0.91*
2	Premenstrual syndrome	0.90*
3	Health status	0.92*

*Significant at 0.01 level of confidence. (Table value required for significance at 0.01 level of confidence is 0.77)

Since the obtained 'R' values were much higher than the required value, the data were accepted as reliable in terms of instrument, tester and the subjects.

3.8 PILOT STUDY

Prior to the formal study sessions, a pilot study was conducted to validate research procedure and the initial capacity of the participants to design the training programme. For the purpose, 10 participants were selected at random, Group I (n=5) underwent intensive interval training and group I (n=5) underwent physical activity intervention for five sessions under the watchful eyes of the investigator. The initial loads of the participants were fixed and the training programme for group I and group II were deigned separately based on the performance in the pilot study. While constructing the training programmes the basic principles of sports training were followed during construction of training programme, the individual differences were also considered.

3.9 TRAINING PROGRAMME

It was a twelve week physical activity intervention programmes for the experimental group. Pre test and post test data were collected from experimental groups and as well as from the control group prior to the commencement of experiments.

The duration of training session in all days was between one hour to one and half hours approximately which included warming up and limbering down. Group II (n=15) acted as control, which did not participate in any specific training on par with experimental groups. All the subjects involved in this study were carefully monitored throughout the training programme to be away from injuries. They were questioned about their health status throughout the training programme. None of them reported with any injuries. However, muscle soreness appeared in the earlier period of the training programme and was reduced in due course.

Group I (n=15) underwent physical activity interventions programme for a duration of 12 weeks and the number of sessions per week was confined to three alternative days. The detailed training programme was presented in Appendix I.

3.10 COLLECTION OF DATA

The data were collected from in Tamilnadu, 1000 students studying bachelors Degree in Tirunelveli were selected at randomly during the academic year 2013 -2014 on selected variables as like Physical activity, premenstrual syndrome, health status and physical activity intervention among the college women students were tested among through standardized test and questionnaire as below.

3.11 ADMINISTRATION OF QUESTIONNAIRE

Level of Physical Activity Questionnaire

The inventory consists of 45 statements. The first ten statements about the subjects of each item, the subject has to select and tick in one of the four columns following: 'strongly agree, agree, disagree, strongly disagree'.

The scores for each item were assigned in the following:

Strongly disagree - 1

Disagree - 2

Agree - 3

Strongly agree - 4

The next ten statements about the subjects of each item, the subject has to select and tick in one of the four columns following: Yes, No

The score for yes is One, Zero other wise.

The next fifteen statements about the subjects of each item, the subject has to select and tick in one of the four columns following: Very important, important, Some what important, No comment.

The scores for each item were assigned in the following:

Very important - 1

Important - 2

Some what important - 3

No comment - 4

3.12 RESEARCH DESIGN AND STATISTICAL PROCEDURE

All the subjects were tested on selected criterion variables, the collection of data from the 1000 women students from various colleges Five women Arts and Science Colleges were selected at random from Tirunelveli districts and these Colleges were Rani Anna College –Tirunelveli, Sarah Tucker College – Tirunelveli, Sri Sarada College – Tirunelveli, Rose Mary College - Tirunelveli and Sri Parasakthi College – Coutralam in Tirunelveli district were statistically analyzed was to examine the descriptive statistics, correlation, dependent'tʼ test data for all variables to determine if any scores were outside possible ranges of health status, The second step consisted thirty women students were selected random as subjects. The data were statistically analyzed by using Analysis of Variance (ANOVA) on Hence, whenever they obtained F- ratio interacted effect was found to be significant, the simple effect test was used as the follow up test.

In all cases 0.05 level 0f significance was fixed and considered to be appropriate in view of the fact that very highly sophisticated equipment were not used for more stringent level of significance.

3.13 SUMMARY

Chapter 3 discussed the methods and materials used to conduct the study. The chapter began with a selection of subjects, variables and tools. The study population and sample were discussed, and the survey instrument was discussed in detail with a discussion of the rationale for choosing selected variables. Finally, data analysis, experimental design and data statistical procedures were discussed.

Chapter IV

ANALYSIS AND INTERPRETATIONS OF THE DATA

4.1 Introduction

The analysis of data and detailed result of the study have been discussed in this chapter. The purpose of the study was to compare the level of physical activity, premenstrual syndrome, health status and to find out the effect of physical activity intervention among the college women students.

To achieve the purpose of the study, 1000 female were selected from the affiliated colleges of Manonmaniam Sundaranar University, Tirunelveli at random. Stratified random sampling method was adopted for this study. Five women's colleges were selected from Tirunelveli districts such as Rani Anna College–Tirunelveli, Sarah Tucker College–Tirunelveli, Sri Saratha College–Tirunelveli, Rose Mary College-Tirunelveli and Sri Parasakthi College-Tirunelveli. Two hundred students were selected at random from each college.

This study consists of two parts. In part I, 1000 students were selected as participants to compare the level of physical activity, premenstrual syndrome and health status by using descriptive statistics, chi-square, t-test and simple ANOVA with Scheffe's post hoc test. In part II, thirty subjects were selected and they were divided in to two groups namely experimental and control groups. Experimental group underwent 12 weeks of exercise intervention programme and control group didn't participated in any specific activity. The effect of exercise intervention on heath status and premenstrual syndrome were analyzed. The data for the study were analyzed using the Statistics Package for the Social Science (SPSS 20.0).

PART-I

One thousand (N=1000) students were selected as participants to analyze the level of physical activity, physical activity barriers, premenstrual syndrome and health status by using descriptive statistics, chi-square, t-test and simple ANOVA.

I. Level of Physical Activity

Table 1: Participation in Yoga / Recreational Activities

Response	Frequency	Percent	Valid Percent
No	520	52.0	52.0
Yes	480	48.0	48.0
Total	1000	100.0	100.0

From the above table, it is observed from the respondents that the percentage score of participating frequently in competitive or recreational sports is 48% and 52% not participating frequently in yoga and recreational activities.

Table 2: Possessing Opinion about Physical activity programs

Response	Frequency	Percent	Valid Percent
No	580	58.0	58.0
Yes	420	42.0	42.0
Total	1000	100.0	100.0

From the above table, it is observed that 58% of our respondents had good opinion/experience with physical activity programs, whereas minimum

42% of respondents had negative opinion/ bad experience about physical activity programs in their life.

Table 3: Willingness to Participate in Sports Competition

Response	Frequency	Percent	Valid Percent
No	520	52.0	52.0
Yes	480	48.0	48.0
Total	1000	100.0	100.0

From the above table, it is observed that 48% of our respondents who were excitedly participated in sports competitions and 52% respondents did not showed their interest in such activities.

Table 4: Participation in Sports activity outside the college

Response	Frequency	Percent	Valid Percent
No	550	55.0	55.0
Yes	450	45.0	45.0
Total	1000	100.0	100.0

From the above table, it is observed that 55% of our respondents have not participated in sports activities outside the college and the remaining 45% respondents shown their interest in participating sports arena outside the college.

Table 5: Symptoms of Injuries Due to Physical Activities

Response	Frequency	Percent	Valid Percent
No	300	30.0	30.0
Yes	700	70.0	70.0
Total	1000	100.0	100.0

From the above table, it is observed that 30% of our respondents who had no back pain injury as a result of involvement in physical activity and minimum of 70% respondents said that they had past experience in back pain as a result of involvement in physical activities.

Table 6: Improve Cardio Vascular Fitness

Response	Frequency	Percent	Valid Percent
No	603	60.3	60.3
Yes	397	39.7	39.7
Total	1000	100.0	100.0

From the above table, it is observed that 60.3% of our respondents have not improved cardio vascular fitness and the remaining 39.7% respondents have improved cardio vascular fitness through physical exercise.

Table 7: Body Fat Weight Loss

Response	Frequency	Percent	Valid Percent
No	693	69.3	69.3
Yes	307	30.7	30.7
Total	1000	100.0	100.0

From the above table, it is observed that 69.3% of our respondents have not reduced body fat/weight and the remaining 39.7% respondents have reduced body fat weight.

Chi-Square test for Class and College teams

Table 8: Class * College Teams Cross tabulation

Class	Values	College_ Teams			Total
		not at all	rarely	very much	
1st year	Count	290	70	40	400
	% within class	72.5%	17.5%	10.0%	100.0%
2 nd year	Count	60	60	220	340
	% within class	17.6%	17.6%	64.7%	100.0%
3 rd year	Count	40	20	200	260
	% within class	15.4%	7.7%	76.9%	100.0%
Total	Count	390	150	460	1000
	% within class	39.0%	15.0%	46.0%	100.0%

From the above table, 76. 9% of the 3rd year students College team is very much interested, 72.5% of First year students the College teams are not at all and 64.7% of second year students the college team is very much interested.

Table 9: Chi-Square Tests for College Teams

Details	Value	df	Asymp. Sig. (2-sided)
Pearson Chi-Square	3.97[a]	4	.000
Likelihood Ratio	438.12	4	.000
Linear-by-Linear Association	325.31	1	.000
N of Valid Cases	1000		

From the above Chi-square test, the significance value is 0.000 which is less than 0.05 level of significance; therefore it was concluded that there is significant association between Class and College Teams. Thus, the 3rd year students shown more interest towards participation in the college sports teams compared to the first and second year students.

Table 10: Residential Area * Physical Activity Cross tabulation

Residential Area	Values	Physical Activity				Total
		0-1	2-3	4-5	5+	
Rural	Count	30	90	360	70	550
	% within present residential area	5.5%	16.4%	65.5%	12.7%	100.0%
Urban	Count	30	90	90	20	230
	% within present residential area	13.0%	39.1%	39.1%	8.7%	100.0%
Town	Count	60	110	40	10	220
	% within present residential area	27.3%	50.0%	18.2%	4.5%	100.0%
Total	Count	120	290	490	100	1000
	% within present residential area	12.0%	29.0%	49.0%	10.0%	100.0%

From the above table, 65.5% of the rural student's participated in physical activity 4 to 5 times in a week, 50 % of students from town area participated in physical activity 2 to 3 times in a week and 39.1 % of the urban students participated in physical activity 4 to 5 times in a week.

Table 11: Chi-Square Tests for Physical Activity

Details	Value	df	Asymp. Sig. (2-sided)
Pearson Chi-Square	2.23	6	.000
Likelihood Ratio	228.05	6	.000
Linear-by-Linear Association	175.28	1	.000
N of Valid Cases	1000		

From the above Chi-square test, the significance value is 0.000 which is less than 0.05 level of significance, it is concluded that there is significant association between Residential area and physical activity. Thus the rural students have very much interested regarding the physical activity compared to the urban and town students.

Level of Physical Activity between Arts and Science Women Students.

Table 12: Independent T-Test between Arts and Science Students on Level of Physical Activity among women college students

Subject	N	Mean	Std. Deviation	Std. Error Mean	t-ratio	df	Sig. (2-tailed)
Arts	502	61.40	21.35	0.95	2.392	998	.017
Science	498	58.29	20.81	0.93			

From the above table, it is observed that the mean values of arts and science students are 61.48 and 58.29 respectively on level of physical activity. The obtained t-test values between arts and science students on level of physical activity is 2.392 which is greater than the p-value (p>0.017). It was conclude that the level of physical activity between arts and science students showed significant difference. However, arts students are better than science students on level of physical activity.

The mean values on level of physical activity between arts and science students were graphically represented in Figure I.

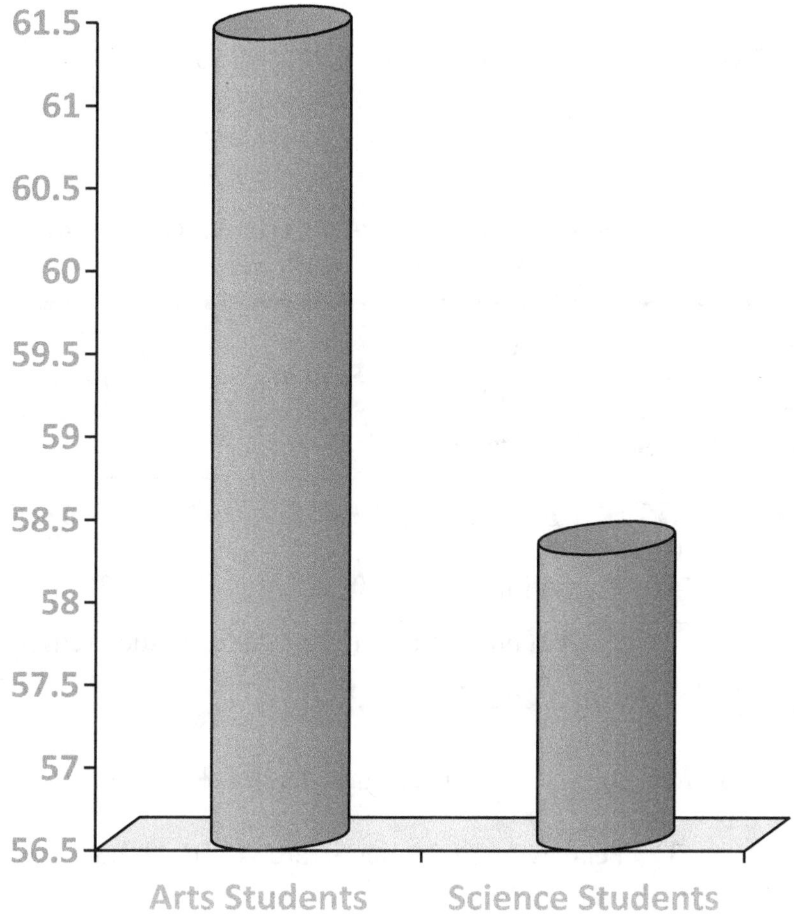

FIGURE 1: MEAN VALUES ON LEVEL OF PHYSICAL ACTIVITY BETWEEN ARTS AND SCIENCE STUDENTS.

Level of Physical Activity among 1st, 2nd & 3rd Year women students

The analysis of variance (ANCOVA) on Level of Physical Activity among 1st, 2nd & 3rd Year women students have been analyzed and presented in table 13.

Table 13 : Analysis of Variance on Level of Physical Activity among 1st, 2nd & 3rd Year women students

Means			Source of Variance	Sum of Squares	df	Mean Squares	'F'-Ratio
1st Year Students	2nd Year Students	3rd Year Students					
43.06 ± 11.81 (N=400)	63.31 ± 19.31 (N=340)	81.35 ± 10.26 (N=260)	Between	236935.14	2	118467.57	**563.97***
			Within	209429.85	997	210.06	

*Significant at .05 level of confidence. (The table value required for significance at 0.05 level with df 2 and 997 is 3.00).

From the above table, the mean values on level of physical activity among 1st, 2nd & 3rd Year women students are 43.06, 63.31 and 81.35 respectively. The obtained F-ratio of 563.97 for means is greater than the table value of 3.00 for df 2 and 997 required for significance at 0.05 level of confidence. Therefore, the results of the study indicated that there was significant difference among 1st, 2nd & 3rd Year women students on level of physical activity. To find out which of the three paired means had a significant difference, the Scheffe's post-hoc test was applied and the results are presented in Table 14.

Table 14 : Scheffe's Test for the Differences among 1st, 2nd & 3rd Year women students on level of physical activity

Mean Values			Mean Differences	Confidence Interval
1st Year Students	2nd Year Students	3rd Year Students		
43.06	63.31	---	20.35*	**2.62**
43.06	--	81.35	40.29*	**2.83**
---	63.31	81.35	18.04*	**2.92**

*Significant at .05 level.

Table above shows that the adjusted post test mean differences on level of physical activity between 1st and 2nd year students; 1st and 3rd year students; and 2nd and 3rd year students are 20.35, 40.29 and 18.04 which are greater than the confidence interval value, which shows significant difference at .05 level of confidence. It may be concluded from the results of the study that there was a significant difference on level of physical activity between 1st and 2nd year students; 1st and 3rd year students; and 2nd and 3rd year students.

It was concluded that 3rd year students is better than 1st and 2nd year students on level of physical activity.

The mean values of on level of physical activity among 1st, 2nd & 3rd Year women students are graphically represented in the Figure II.

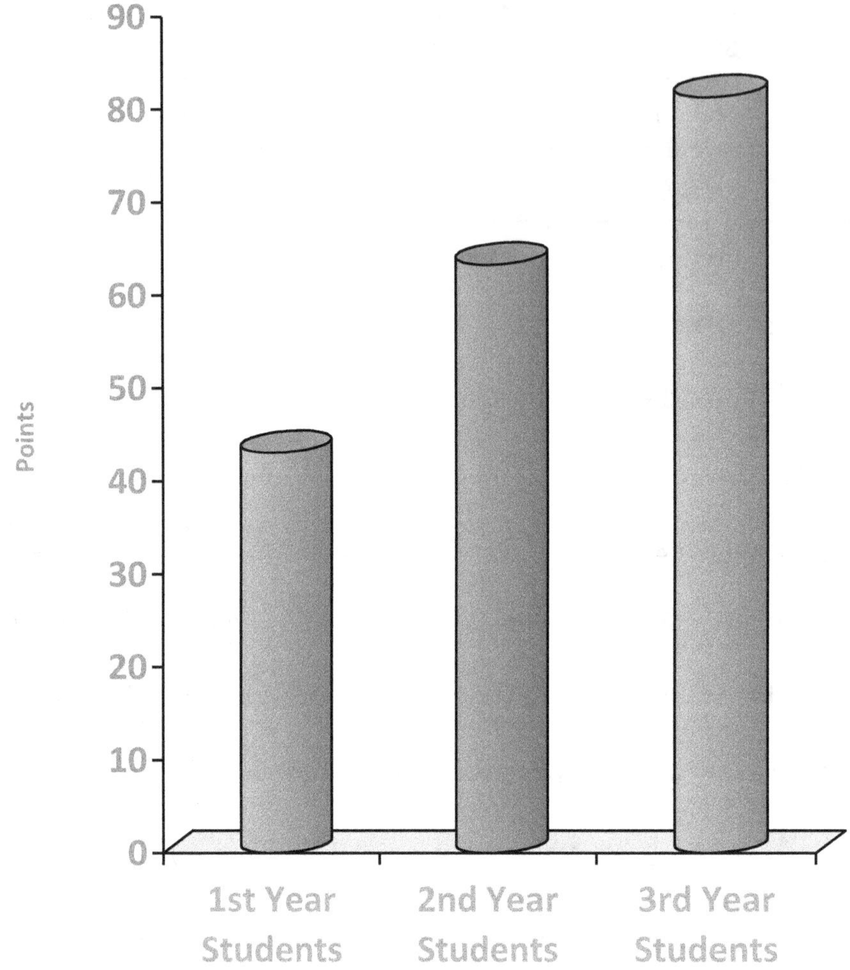

FIGURE II: MEAN VALUES OF ON LEVEL OF PHYSICAL ACTIVITY AMONG 1ST, 2ND & 3RD YEAR WOMEN STUDENTS.

11. Health Status

Table 15: Fever/Chills/Sweats

Response	Frequency	Percent	Valid Percent
No	670	67.0	67.0
Yes	330	33.0	33.0
Total	1000	100.0	100.0

From the above table, it is observed that 67% of our respondents had no fever/chills/sweats involving physical activity programs, whereas minimum 33% of respondents had fever/chills/sweats involving physical activity programs in their life.

Table 16: Shortness of Breath/Cough

Response	Frequency	Percent	Valid Percent
No	640	64.0	64.0
Yes	360	36.0	36.0
Total	1000	100.0	100.0

From the above table, it is observed that 64% of our respondents had no Shortness of Breath/Cough involving physical activity programs, whereas 36% of respondents had Shortness of Breath/Cough involving physical activity programs in their life.

Table 17: Dizziness or Loss of Consciousness

Response	Frequency	Percent	Valid Percent
No	530	53.0	53.0
Yes	470	47.0	47.0
Total	1000	100.0	100.0

From the above table, it is observed that 53% of our respondents had no Loss of Consciousness involving physical activity programs, whereas 47% of respondents had Loss of Consciousness involving physical activity programs in their life.

Table 18: Trouble with your period

Response	Frequency	Percent	Valid Percent
No	560	56.0	56.0
Yes	440	44.0	44.0
Total	1000	100.0	100.0

From the above table, it is observed that 44% of our respondents had no trouble with involving physical activity programs, whereas 56% of respondents had trouble with involving physical activity programs in their life.

Table 19: Nausea/Vomiting

Response	Frequency	Percent	Valid Percent
no	650	65.0	65.0
yes	350	35.0	35.0
Total	1000	100.0	100.0

From the above table, it is observed that 65% of our respondents had no Nausea/Vomiting with involving physical activity programs, whereas 35% of respondents had Nausea/Vomiting with involving physical activity programs in their life.

Health Status between Arts and Science Students

In order to compare the Subject based on the percentage score for health status, cross tabulation is given below:

Table 20: Subject * Health Status - Cross Tabulation

Subject	Details	Health Status					Total
		very poor	Poor	average	good	very good	
Arts	Count	60	60	98	124	160	502
	% within subject	11.95%	11.95%	19.52%	24.70%	31.87%	100.0%
Science	Count	90	200	118	30	60	498
	% within subject	18.07%	40.16%	23.69%	6.02%	12.05%	100.0%
Total	Count	150	260	216	154	220	1000
	% within subject	15.0%	26.0%	21.6%	15.4%	22.0%	100.0%

From the above table, it is found that 31.87% of Arts Students having very good health status, 40.16% of Science Students has poor health status.

Table 21: Chi-Square Tests for Health Status

Details	Value	df	Asymp. Sig. (2-sided)
Pearson Chi-Square	4.172E2[a]	12	.019
Likelihood Ratio	482.180	12	.011
Linear-by-Linear Association	3.486	1	.062
N of Valid Cases	1000		

From the above Chi-square test, the significance value is 0.019 which is less than 0.05 level of significance, it is concluded that there is significant association between health status and subjects. Thus the science students had poor health status compared to the arts students.

Health status and Residential Area

Table 22: Residential Area * Disorder Cross tabulation

Residential Area	Details	Health Status					
		very severe	severe	moderate	mild	very mild	Total
Urban	Count	75	60	70	10	15	230
	% within residential_area	32.61%	26.09%	30.43%	4.35%	6.52%	100.0%
Town	Count	40	91	54	19	16	220
	% within residential_area	18.18%	41.36%	24.55%	8.64%	7.27%	100.0%
Rural	Count	25	30	155	175	165	550
	% within residential_area	4.55%	5.45%	28.18%	31.82%	30%	100.0%
Total	Count	140	181	279	204	196	1000
	% within residential_area	14%	18.1%	27.9%	20.4%	19.6%	100.0%

From the above table, 32.61% of urban Students had very severe status, 41.36% of town students had severe health status and 27.9% of village students had moderate health status.

Table 23: Chi-Square Tests for health status and Residential Area

Details	Value	df	Asymp. Sig. (2-sided)
Pearson Chi-Square	5.820E2[a]	8	.020
Likelihood Ratio	677.107	8	.420
Linear-by-Linear Association	358.208	1	.720
N of Valid Cases	1000		

From the above table, it is observed that the significance 0.020 which is less than 0.05 level of significance, therefore it is concluded that there is significant difference in health status and residential areas. Thus the urban students have severe health status compared to the town and rural area students.

III. Premenstrual Syndrome (PMS)

PMS between Arts and Science Students

Table 24: Independent T-Test between Arts and Science Students on Premenstrual syndrome among women college students

Subject	N	Mean	Std. Deviation	Std. Error Mean	t-ratio	df	Sig. (2-tailed)
Arts	502	56.01	16.88	.75	2.10	998	.035
Science	498	53.85	15.52	.69			

From the above table, it is observed that the mean values of arts and science students are 61.48 and 58.29 respectively on premenstrual syndrome. The obtained t-test values between arts and science students on premenstrual syndrome is 2.106 which is greater than the p-value (p>0.035). It was concluded that the premenstrual syndrome between arts and science students showed significant difference. However, arts students are better than science students on premenstrual syndrome.

The mean values on premenstrual syndrome between arts and science students were graphically represented in Figure I.

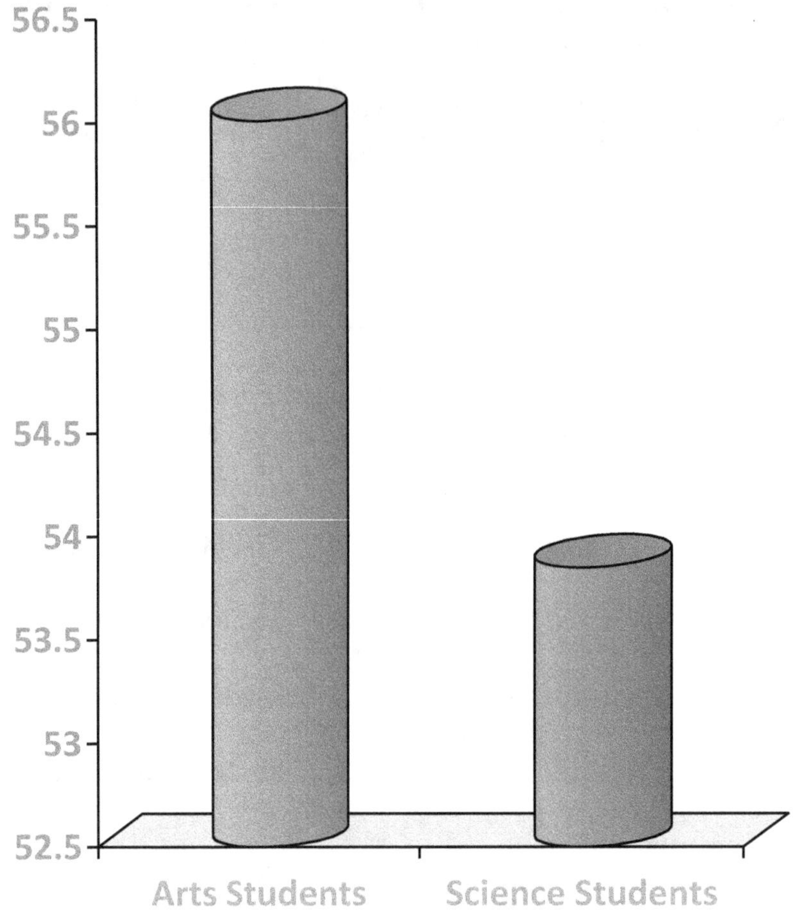

FIGURE III: MEAN VALUES ON PREMENSTRUAL SYNDROME BETWEEN ARTS AND SCIENCE STUDENTS.

PMS among Urban, Town and Rural College Women Students

The analysis of variance (ANOVA) on premenstrual syndrome among 1st, 2nd & 3rd Year women students have been analyzed and presented in table 13.

Table 25 : Analysis of Variance on premenstrual syndrome among 1st, 2nd & 3rd Year women students

Means			Source of Variance	Sum of Squares	df	Mean Squares	'F'-Ratio
Urban	Town	Rural					
51.39 ± 9.56 (N=230)	56.39 ± 9.25 (N=220)	53.92 ± 11.30 (N=550)	Between	2402.63	2	1201.32	**11.60***
			Within	103222.31	997	103.53	

*Significant at .05 level of confidence. (The table value required for significance at 0.05 level with df 2 and 997 is 3.00).

From the above table, the mean values on premenstrual syndrome among urban, town and rural college women students are 51.39, 56.39 and 53.92 respectively. The obtained F-ratio of 11.60 for means is greater than the table value of 3.00 for df 2 and 997 required for significance at 0.05 level of confidence. Therefore, the results of the study indicated that there was significant difference on premenstrual syndrome among urban, town and rural college women students. To find out which of the three paired means had a significant difference, the Scheffe's post-hoc test was applied and the results are presented in Table 14.

Table 26 : Scheffe's Test for the Differences on premenstrual syndrome among urban, town and rural college women students

Mean Values			Mean Differences	Confidence Interval
1st Year Students	2nd Year Students	3rd Year Students		
51.39	56.39	---	5.00*	**2.35**
51.39	--	53.92	2.53*	**1.96**
---	56.39	53.92	2.47*	**1.99**

*Significant at .05 level.

Table above shows that the adjusted post test mean differences on premenstrual syndrome between urban and town college women students; urban and rural college women students; and town and rural college women students are 5.00, 2.53 and 2.47 which are greater than the confidence interval value, which shows significant difference at .05 level of confidence. It may be concluded from the results of the study that there was a significant difference on premenstrual syndrome between urban and town college women students; urban and rural college women students; and town and rural college women students on premenstrual syndrome.

It was concluded that rural students is better than urban and town area college women students on premenstrual syndrome.

The mean values of on premenstrual syndrome among urban, town and rural area college women students are graphically represented in the Figure IV.

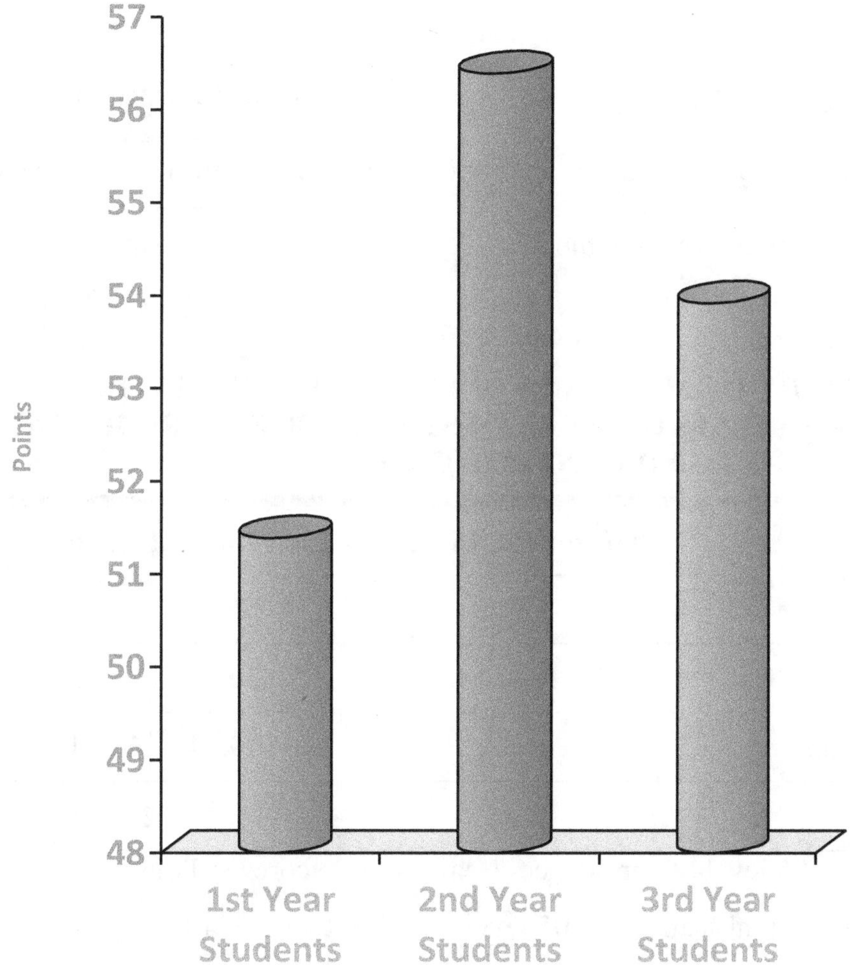

FIGURE IV: MEAN VALUES OF ON PREMENSTRUAL SYNDROME AMONG URBAN, TOWN AND RURAL COLLEGE WOMEN STUDENTS.

PART II

The effect of exercise intervention programme on heath status and premenstrual syndrome were analyzed. The data for the study were analyzed using the Statistics Package for Social Science (SPSS 20.0).

Health Status

The analysis of dependent 't'-test on the data obtained for health status of the pre-test and post-test means of Experimental and control groups have been analyzed and presented in table 27.

TABLE 27

THE SUMMARY OF MEAN AND DEPENDENT 't' TEST FOR THE PRE AND POST TESTS ON HEALTH STATUS OF EXPERIMENTAL AND CONTROL GROUPS

Mean	**Experimental Group**		**Control Group**	
	Mean	SD	Mean	SD
Pre test Mean	32.31	2.34	30.31	3.56
Post test Mean	17.54	3.17	29.34	4.11
't' test	**14.21***		**1.25**	

*Significant at 0.05 level of confidence. Health status scores in Points. (Table value required for significance at 0.05 level for 't' test with df 14 is 2.14).

From the table above, the dependent 't'-test values between the pre and post tests means of Experimental and control groups were 14.21 and 1.25 respectively. Since, the obtained 't'-test value of experimental group was greater than the table value 2.14 with df 14 at 0.05 level of confidence, it is concluded that experimental group had significant improvement of the participants' health status. However, control group has no significant improvement in health status because they were not subjected to any specific training other than their regular routine.

The pre and post test mean values of experimental and control groups on health status are graphically represented in the figure V.

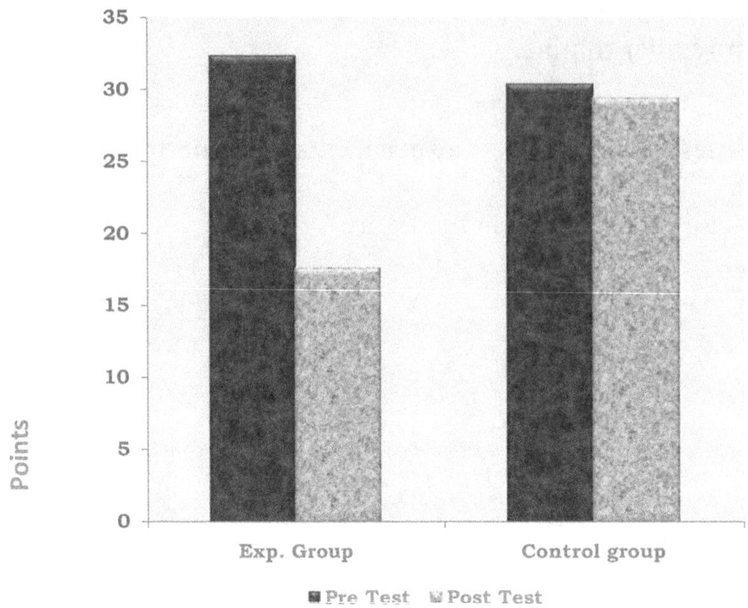

FIGURE V: PRE AND POST TEST MEAN VALUES OF EXPERIMENTAL AND CONTROL GROUPS ON HEALTH STATUS.

The analysis of covariance (ANCOVA) on health status of Experimental and control groups have been analyzed and presented in table 28.

TABLE 28

ANALYSIS OF COVARIANCE OF EXPERIMENTAL AND CONTROL GROUPS ON HEALTH STATUS

Adjusted Post Test Means		Source of Variance	Sum of Squares	df	Mean Squares	'F'-Ratio
Experimental Group	**Control Group**					
17.41	30.31	Between	214.23	1	214.23	**17.79***
		Within	325.14	27	12.04	

*Significant at .05 level of confidence. (The table value required for significance at 0.05 level with df 1 and 27 is 3.35).

From the above table, the adjusted post test mean values of health status for experimental and control groups are 17.41 and 30.31 respectively. The obtained F-ratio of 17.79 for adjusted post test means is greater than the table value of 3.35 for df 1 and 27 required for significance at 0.05 level of confidence. Therefore, the results of the study indicated that there was significant difference between the adjusted post test means of experimental and control groups on the health status of participants. Experimental group had effect on participants' health status.

The adjusted post test mean values of experimental and control groups on health status are graphically represented in the figure VI.

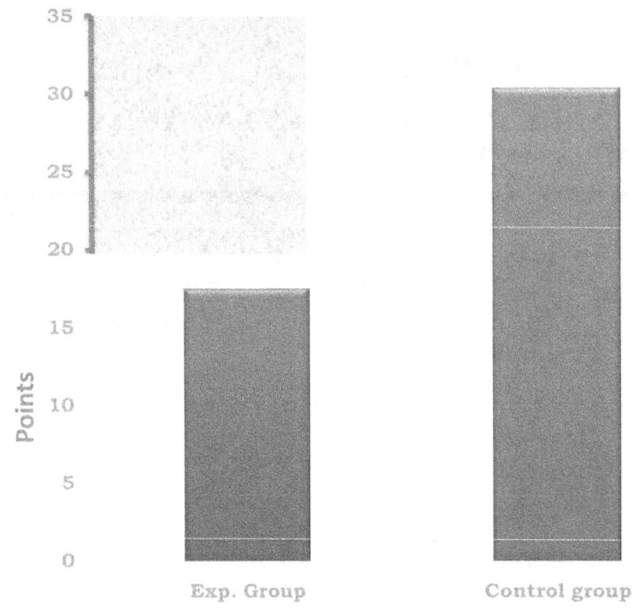

FIGURE VI: THE ADJUSTED POST TEST MEAN VALUES OF EXPERIMENTAL AND CONTROL GROUP ON HEALTH STATUS.

Premenstrual Syndrome

The analysis of dependent 't'-test on the data obtained for Premenstrual Syndrome of the pre-test and post-test means of Experimental and control groups have been analyzed and presented in table 27.

TABLE 29

THE SUMMARY OF MEAN AND DEPENDENT 't' TEST FOR THE PRE AND POST TESTS ON PREMENSTRUAL SYNDROME OF EXPERIMENTAL AND CONTROL GROUPS

Mean	Experimental Group		Control Group	
	Mean	SD	Mean	SD
Pre test Mean	54.32	3.56	55.21	3.98
Post test Mean	36.45	4.12	54.97	3.78
't' test	**11.24***		**0.87**	

*Significant at 0.05 level of confidence. Premenstrual Syndrome scores in Points. (Table value required for significance at 0.05 level for t_{14} is 2.14).

From the table above, the dependent 't'-test values between the pre and post tests means of Experimental and control groups were 11.24 and 0.87 respectively. Since, the obtained 't'-test value of experimental group was greater than the table value 2.14 with df 14 at 0.05 level of confidence, it is concluded that experimental group had significant improvement of the participants' Premenstrual Syndrome. However, control group has no significant improvement in health status because they were not subjected to any specific training other than their regular routine.

The pre and post test mean values of experimental and control groups on Premenstrual Syndrome are graphically represented in the figure VII.

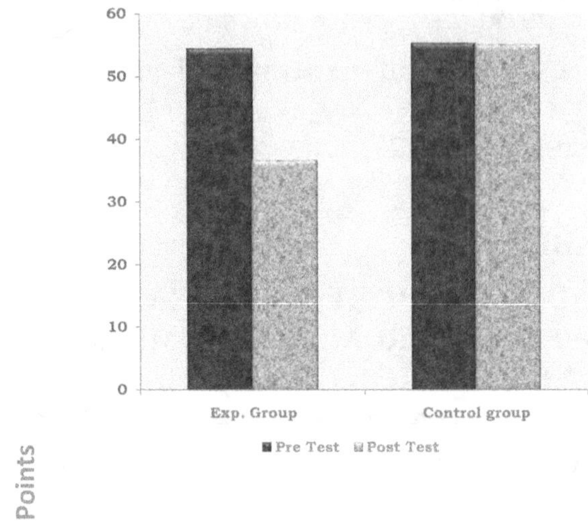

FIGURE VII: PRE AND POST TEST MEAN VALUES OF EXPERIMENTAL AND CONTROL GROUPS ON PREMENSTRUAL SYNDROME.

The analysis of covariance (ANCOVA) on Premenstrual Syndrome of Experimental and control groups have been analyzed and presented in table 28.

TABLE 30

ANALYSIS OF COVARIANCE OF EXPERIMENTAL AND CONTROL GROUPS ON PREMENSTRUAL SYNDROME

Adjusted Post Test Means		Source of Variance	Sum of Squares	df	Mean Squares	'F'-Ratio
Experimental Group	Control Group					
36.80	55.11	Between	424.51	1	424.51	**15.38***
		Within	745.21	27	27.60	

*Significant at .05 level of confidence. (The table value required for significance at 0.05 level with df 1 and 27 is 3.35).

From the above table, the adjusted post test mean values of Premenstrual Syndrome for experimental and control groups are 36.80 and 55.11 respectively. The obtained F-ratio of 15.38 for adjusted post test means is greater than the table value of 3.35 for df 1 and 27 required for significance at 0.05 level of confidence. Therefore, the results of the study indicated that there was significant difference between the adjusted post test means of experimental and control groups on the Premenstrual Syndrome of participants. Experimental group had effect on participants' Premenstrual Syndrome.

The adjusted post test mean values of experimental and control groups on Premenstrual Syndrome are graphically represented in the figure VIII.

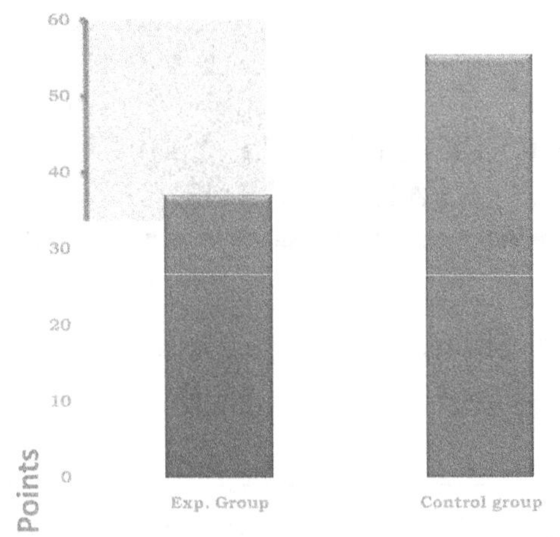

FIGURE VIII: THE ADJUSTED POST TEST MEAN VALUES OF EXPERIMENTAL AND CONTROL GROUP ON PREMENSTRUAL SYNDROME.

4.2 Discussion on Findings

Findings for Physical Activity Level

- Subject to college teams and class, college teams for the 3rd year students have very much interested when compared to other students.
- Subject to physical activity level and residential area, physical activity levels for the rural students have better than other students.
- Subject to skipping and class, the skipping for 3rd year students has better than other students.
- With reference to the subjects, physical activity level for arts students is better than other students.
- With reference to the class, physical activity level for 3rd year students is better than other students.
- With reference to the colleges, physical activity level for Rose Mary College students is better than other college students.
- With reference to the colleges, daily physical activity level for Rose Mary College students is better than other college students.
- With reference to the residential area, participating physical activity for arts students is higher than other students.
- With reference to the colleges, importance of physical activity for Rose Mary College students is better than other college students.
- There is no significant difference in the time spending for moderate physical activity score between the subjects.
- With reference to colleges, moderate physical activity for Sri Parasakthi College students is very much interested than other students.
- With reference to subjects, time spending for vigorous physical activity for arts students is better than other students.
- With reference to colleges, time spending for vigorous physical activity for Sri Parasakthi College students is better than other college students.

4.3 Findings for Health Status

- Subject to health status and subjects, the management students have poor health status when compared to other students.
- Subject to disorder and residential area, the town students have severe disorder than others.
- Subject to headache and academic status, PG students have severe headache than UG students.
- With reference to academic category, health status for UG students is better than PG students.
- With reference to academic category, breathing status for UG students is better than PG students.
- With reference to subjects, breathing status for arts students is better than other subjects.
- With reference to academic status, UG students have less body pain compared to the PG students.
- With reference to residential area, Village students have less body pain compared to the other residence.
- With reference to academic category, mental health for UG students is better than PG students.
- With reference to colleges, mental health for Sarah Tucker College students have better than other college students.
- With reference to the academic status, physical sate for UG students is better than PG students.
- With reference to colleges, physical state for Sarah Tucker College students is better than other college students.
- With reference to academic status, PMS_A for UG students have less premenstrual symptoms than PG Students.
- With reference to the subjects, PMS_A for arts students have less premenstrual symptoms than other subject Students.

- There is no significant difference in PMS_B score between academic status.
- With reference to the areas of residence, PMS_B for town students have less premenstrual symptoms than others.
- With reference to the academic status, PMS_C for UG students have less premenstrual symptoms than PG Students.
- With reference to the areas of residence, PMS_C for village students have less premenstrual symptoms than others.

4.4 Findings for Experimental and Control Group

- With reference to pre and post test physical activity levels for post activity level is better than pre test physical activity levels.
- With reference to pre and post test physical activity levels, there is low correlation between pre and post test for physical activity levels.
- There is no significant difference between physical activity levels (pre and post test) among college levels.
- With reference to pre and post test participating physical activity for post activity level is better than pre test participating physical activity.
- With reference to pre and post test participating physical activity, there is low correlation between pre and post test for participating physical activity.
- There is no significant difference between participating physical activity (pre and post test) among college levels.
- With reference to pre and post test moderate physical activity for post moderate physical activity is better than pre test moderate physical activity.
- With reference to pre and post test moderate physical activity, there is low correlation between pre and post test for moderate physical activity.
- There is no significant difference between moderate physical activity (pre and post test) among college levels.

- With reference to pre and post test vigorous physical activity for post vigorous physical activity is better than pre test vigorous physical activity.
- With reference to pre and post test vigorous physical activity, there is low correlation between pre and post test for vigorous physical activity.
- There is no significant difference between vigorous physical activity (pre and post test) among college levels.
- With reference to pre and post test physical state for post physical state is better than pre test physical state.
- With reference to pre and post test physical state, there is low correlation between pre and post test for physical state.
- There is no significant difference between physical state (pre and post test) among college levels.
- With reference to pre and post test mental health for post mental health is better than pre test mental health.
- With reference to pre and post test mental health, there is low correlation between pre and post test for mental health.
- There is no significant difference between mental health (pre and post test) among college levels.
- With reference to pre and post test PMS_A for post PMS_A is better than pre test PMS_A.
- With reference to pre and post test PMS_A, there is low correlation between pre and post test for PMS_A.
- There is no significant difference between PMS_A (pre and post test) among college levels.
- With reference to pre and post test PMS_B for post PMS_B is better than pre test PMS_B.
- With reference to pre and post test PMS_B, there is low correlation between pre and post test for PMS_B.

- There is no significant difference between PMS_B (pre and post test) among college levels.
- With reference to pre and post test PMS_C for post PMS_C is better than pre test PMS_C.
- With reference to pre and post test PMS_C, there is low correlation between pre and post test for PMS_C.
- There is no significant difference between PMS_C (pre and post test) among college levels.

4.5 Discussion on Hypothesis

1. It was hypothesized in the beginning that there would be significant difference on level of physical activity, premenstrual syndrome and health status among college level women students. The results of the study produced significant results, Hence, the researcher first hypothesis was accepted and null hypothesis was rejected.
2. There would be significant difference on level of physical activity, premenstrual syndrome and health status among college level women students. The results of the study produced significant results, Hence, the researcher second hypothesis was accepted and null hypothesis was rejected.
3. There would be significant improvement on premenstrual syndrome and health status due to the effects of physical activity intervention. The results of the study produced significant results, Hence, the researcher third hypothesis was accepted and null hypothesis was rejected.
4. There would be significant difference between experimental and control groups between experimental and control groups. The results of the study produced significant results, Hence, the researcher fourth hypothesis was accepted and null hypothesis was rejected.

Chapter V

SUMMARY, CONCLUSIONS AND RECOMMENDATIONS

5.1 Summary

The purpose of the study was to compare the level of physical activity, premenstrual syndrome, health status and to find out the effect of physical activity intervention among the college women students. To achieve the purpose of the study, 1000 female were selected from the affiliated colleges of Manonmaniam Sundaranar University, Tirunelveli at random. Stratified random sampling method was adopted for this study. Five women's colleges were selected from Tirunelveli districts such as Rani Anna College–Tirunelveli, Sarah Tucker College–Tirunelveli, Sri Saratha College–Tirunelveli, Rose Mary College-Tirunelveli and Sri Parasakthi College-Tirunelveli. Two hundred students were selected at random from each college.

This study consists of two parts. In part I, 1000 students were selected as participants to compare the level of physical activity, premenstrual syndrome and health status by using descriptive statistics, chi-square, t-test and simple ANOVA with Scheffe's post hoc test. In part II, thirty subjects were selected and they were divided in to two groups namely experimental and control groups. Experimental group underwent 12 weeks of exercise intervention programme and control group didn't participated in any specific activity. The effect of exercise intervention on heath status and premenstrual syndrome were analyzed. The data for the study were analyzed using the Statistics Package for the Social Science (SPSS 20.0).

The variables selected for this were physical activity level, health status and premenstrual symptoms. The selected variables were tested with standardized questionnaire. The data collected from the subjects were statistically examined for Chi Square, Descriptive Statistics, independent t test, dependent t test, ANOVA, Scheffe's Test, ANCOVA test. In this case, the criterion for statistical significance would set as 0.05 level of confidence.

5.2 Conclusions

- It is highly evident that the students ensures good health after post test than pre test
- Physical activity was a part of the human daily life. It has been significantly improved the students life style.
- Physical activity appears to relieve symptoms of depression and anxiety and improve mood.
- There is promising evidence that strength training and other forms of exercise in students preserve the ability to maintain independent living status and reduce the risk of falling.
- Physical activity feels the students to feel better and get more energy, happy and relaxed, and sleep better.
- This study provides valuable research to the growing body of knowledge on physical activity, health status and premenstrual symptoms.

- o There is strong support from the results after training the students, can help individuals develop healthy habits of physical activity and premenstrual symptoms.
- o Experimental group and control group were compared. Experimental group better than control group.

5.3 Recommendations

- ➢ It is recommended that further research be designed to investigate the effects of physical activity on menstrual symptoms.
- ➢ It is recommended that further research be designed to investigate with an elite subject population.
- ➢ It is recommended that further research be designed to investigate with previously endurance and resistance trained subjects
- ➢ It is recommended that further research be conducted using more subjects.

BIBLIOGRAPHY

Ahangar, K. S., Dolatshahei, B., & Badeie, M. M. (2011). Efficacy of Anger Management Training on Anger Severity Symptoms of Premenstrual Syndrome. *Procedia-Social and Behavioral Sciences*, *30*, 1942-1944.

Al-Isa, A. N., Campbell, J., Desapriya, E., & Wijesinghe, N. (2011). Social and health factors associated with physical activity among Kuwaiti college students. *Journal of obesity*, *2011*.

Al-Kandari, F., & Vidal, V. L. (2007). Correlation of the health-promoting lifestyle, enrollment level, and academic performance of College of Nursing students in Kuwait. *Nursing & health sciences*, *9*(2), 112-119.

Allais, G., Gabellari, I. C., Burzio, C., Rolando, S., De Lorenzo, C., Mana, O., & Benedetto, C. (2012). Premenstrual syndrome and migraine. *Neurological Sciences*, *33*(1), 111-115.

Allen, L.M., Lam, A.C. (2012, Apr) Premenstrual syndrome and dysmenorrhea in adolescents. *Adolesc Med State Art Rev.* 2012 Apr; 23(1):139-63.

Allender, S., Cowburn, G., & Foster, C. (2006). Understanding participation in sport and physical activity among children and adults: a review of qualitative studies. *Health education research*, *21*(6), 826-835.

Andegiorgish, A. K., Wang, J., Zhang, X., Liu, X., & Zhu, H. (2012). Prevalence of overweight, obesity, and associated risk factors among school children and adolescents in Tianjin, China. *European journal of pediatrics*, *171*(4), 697-703.

Andrade Osorio, É., Padilla Raygoza, N., & Ruiz Paloalto, M. (2013). Perceived barriers and physical activity level in older adults from Aguascalientes, Ags.: un studio transversal. *Enfermería Global*, (31), 43.

Arbabi, M., Shirmohammadi, M., Taghizadeh, Z., Haghanni, H., & Parsafar, H. (2011). P01-368-Premenstrual syndrome and psychiatric comorbidities. *European Psychiatry*, *26*, 370.

Bale, J. M., Gazmararian, J. A., & Elon, L. (2015). Effect of the work environment on using time at work to exercise. *American Journal of Health Promotion*, *29*(6), 345-352.

Barnes, P. M., & Schoenborn, C. A. (2003). *Physical activity among adults, United States, 2000*. US Department of Health and Human Services, Centers for Disease Control and Prevention, National Center for Health Statistics.

Bastani, F., & Hashemi, S. (2012). Effects of a web based lifestyle education on general health and severity of the symptoms of Premenstrual Syndrome (PMS) among female students: a randomized controlled trial. *Procedia-Social and Behavioral Sciences*, *46*, 3356-3362.

Basu, S., Seligman, H., & Winkleby, M. (2014). A metabolic–epidemiological microsimulation model to estimate the changes in energy intake and physical activity necessary to meet the Healthy People 2020 obesity objective. *American journal of public health*, *104*(7), 1209-1216.

Begg, S. J., Vos, T., Barker, B., Stanley, L., & Lopez, A. D. (2008). Burden of disease and injury in Australia in the new millennium: measuring health loss from diseases, injuries and risk factors. *Medical journal of Australia*, *188*(1), 36.

Bellew, B., Schöeppe, S., Bull, F. C., & Bauman, A. (2008). The rise and fall of Australian physical activity policy 1996–2006: a national review framed in an international context. *Australia and New Zealand health policy*, *5*(1), 18.

Bendich, A. (2001). Micronutrients in women's health and immune function. *Nutrition*, *17*(10), 858-867.

Bergenthal, N., Will, A., Streckmann, F., Wolkewitz, K. D., Monsef, I., Engert, A., ... & Skoetz, N. (2014). Aerobic physical exercise for adult patients with haematological malignancies. *The Cochrane Library*.

Berger, G., & Peerson, A. (2009). Giving young Emirati women a voice: participatory action research on physical activity. *Health & place*, *15*(1), 117-124.

Bertone-Johnson, E. R., Chocano-Bedoya, P. O., Zagarins, S. E., Micka, A. E., & Ronnenberg, A. G. (2010). Dietary vitamin D intake, 25-hydroxyvitamin D 3 levels and premenstrual syndrome in a college-aged population. *The Journal of steroid biochemistry and molecular biology*, *121*(1), 434-437.

Bertone-Johnson, E. R., Hankinson, S. E., Bendich, A., Johnson, S. R., Willett, W. C., & Manson, J. E. (2005). Calcium and vitamin D intake and risk of incident premenstrual syndrome. *Archives of internal medicine*, *165*(11), 1246-1252.

Better health retrieves from http://www.betterhealth.vic.gov.au/bhcv2/bhcarticles.nsf/pages/Physical_activity_women?open (April 2014)

Better health retrieves from http://www.betterhealth.vic.gov.au/bhcv2/bhcarticles.nsf/pages/Physical_activity_women?open (April 2014)

Better health retrieves from http://www.betterhealth.vic.gov.au/bhcv2/bhcarticles.nsf/pages/premenstrual_syndrome?open (May 2014)

Better health retrieves from http://www.betterhealthchannel.com.au/bhcv2/bhcarticles.nsf/pages/Teenage_girls_getting_them_active?open

Biggs, W. S., & Demuth, R. H. (2011). Premenstrual syndrome and premenstrual dysphoric disorder. *American family physician*, *84*(8).

Blamey, A., & Mutrie, N. (2004). Changing the individual to promote health-enhancing physical activity: the difficulties of producing evidence and translating it into practice. *Journal of sports sciences*, *22*(8), 741-754.

Bloom, D. E., Canning, D., & Sevilla, J. (2004). The effect of health on economic growth: a production function approach. *World development*, *32*(1), 1-13.

Blumel, J. E., Castelo-Branco, C., Binfa, L., Gramegna, G., Tacla, X., Aracena, B., ... & Sanjuan, A. (2000). Quality of life after the menopause: a population study. *Maturitas*, *34*(1), 17-23.

Bodde, A. E., Seo, D. C., & Frey, G. (2009). Correlation between physical activity and self-rated health status of non-elderly adults with disabilities. *Preventive medicine*, *49*(6), 511-514.

Booth, F. W., Laye, M. J., Lees, S. J., Rector, R. S., & Thyfault, J. P. (2008). Reduced physical activity and risk of chronic disease: the biology behind the consequences. *European journal of applied physiology*, *102*(4), 381-390.

Borenstein, J. E., Dean, B. B., Leifke, E., Korner, P., & Yonkers, K. A. (2007). Differences in symptom scores and health outcomes in premenstrual syndrome. *Journal of Women's Health*, *16*(8), 1139-1144.

Boyle, J., Mattern, C. O., Lassiter, J. W., & Ritzler, J. A. (2011). Peer 2 peer: efficacy of a course-based peer education intervention to increase physical activity among college students. *Journal of American College Health*, *59*(6), 519-529.

Boyle, S. E., Jones, G. L., & Walters, S. J. (2008). Physical activity among adolescents and barriers to delivering physical education in Cornwall and Lancashire, UK: a qualitative study of heads of PE and heads of schools. *BMC public health*, *8*(1), 273.

Boyle, S. E., Jones, G. L., & Walters, S. J. (2008). Physical activity among adolescents and barriers to delivering physical education in Cornwall and Lancashire, UK: a qualitative study of heads of PE and heads of schools. *BMC public health*, *8*(1), 273.

Bravata, D. M., Smith-Spangler, C., Sundaram, V., Gienger, A. L., Lin, N., Lewis, R., ... & Sirard, J. R. (2007). Using pedometers to increase physical activity and improve health: a systematic review. *Jama*, *298*(19), 2296-2304.

Breslow, R. A., Ballard-Barbash, R., Munoz, K., & Graubard, B. I. (2001). Long-term recreational physical activity and breast cancer in the National Health and

Nutrition Examination Survey I epidemiologic follow-up study. *Cancer Epidemiology and Prevention Biomarkers*, *10*(7), 805-808.

Brooker, K., Van Dooren, K., McPherson, L., Lennox, N., & Ware, R. (2015). Systematic Review of Interventions Aiming to Improve Involvement in Physical Activity Among Adults With Intellectual Disability. *Journal of Physical Activity and Health*, *12*(3), 434-444.

Brosse, A. L., Sheets, E. S., Lett, H. S., & Blumenthal, J. A. (2002). Exercise and the treatment of clinical depression in adults. *Sports medicine*, *32*(12), 741-760.

Brown, J. K., Byers, T., Doyle, C., Courneya, K. S., Demark-Wahnefried, W., Kushi, L. H., ... & Eldridge, B. (2003). Nutrition and physical activity during and after cancer treatment: an American Cancer Society guide for informed choices. *CA: a cancer journal for clinicians*, *53*(5), 268-291.

Brown, S. A. (2005). Measuring perceived benefits and perceived barriers for physical activity. *American journal of health behavior*, *29*(2), 107-116.

Brown, S. A., Huber, D., & Bergman, A. (2006). A perceived benefits and barriers scale for strenuous physical activity in college students. *American Journal of Health Promotion*, *21*(2), 137-140.

Brown, S. C., Geiselman, P. J., Copeland, A. L., Carol Gordon RN, P. H. D., & Richard-Eaglin, A. (2008). Reliability and validity of the Personal Wellness Profile (PWP) Questionnaire in African American college women. *Journal of cultural diversity*, *15*(4), 163.

Buffart, L. M., Thong, M. S., Schep, G., Chinapaw, M. J., Brug, J., & Van De Poll-franse, L. V. (2012). Self-reported physical activity: its correlates and relationship with health-related quality of life in a large cohort of colorectal cancer survivors. *PloS one*, *7*(5), e36164.

Bula-Biteniece, I., & Grants, J. (2010). Students' professional training for organizing family sport: Professional training and family sport. *Polish Journal of Sport and Tourism, 17 (4)*, 245-247.

Bull, F. C., & Bauman, A. E. (2011). Physical inactivity: the "Cinderella" risk factor for noncommunicable disease prevention. *Journal of health communication*, *16*(sup2), 13-26.

Bull, F. C., Armstrong, T. P., DixON, T., Ham, S., NEiMAN, A. N. D. R. E. A., & Pratt, M. (2004). Physical inactivity. *Comparative Quantification of Health Risks Global and Regional Burden of Disease Attributable to Selected Major Risk Factors. Geneva: World Health Organization*, 729-881.

Busse, J. W., Montori, V. M., Krasnik, C., Patelis-Siotis, I., & Guyatt, G. H. (2009). Psychological intervention for premenstrual syndrome: a meta-analysis of randomized controlled trials. *Psychotherapy and psychosomatics*, *78*(1), 6-15.

Campagne, D. M., & Campagne, G. (2007). The premenstrual syndrome revisited. *European Journal of obstetrics & Gynecology and reproductive biology*, *130*(1), 4-17.

Canadell, J. G., Le Quéré, C., Raupach, M. R., Field, C. B., Buitenhuis, E. T., Ciais, P., ... & Marland, G. (2007). Contributions to accelerating atmospheric

CO_2 growth from economic activity, carbon intensity, and efficiency of natural sinks. *Proceedings of the national academy of sciences*, *104*(47), 18866-18870.

Carlson, J. J., Eisenmann, J. C., Pfeiffer, K. A., Jager, K. B., Sehnert, S. T., Yee, K. E., ... & Feltz, D. L. (2008). (S) Partners for Heart Health: a school-based program for enhancing physical activity and nutrition to promote cardiovascular health in 5 th grade students. *BMC public health*, *8*(1), 420.

Carter, R., Vos, T., Moodie, M., Haby, M., Magnus, A., & Mihalopoulos, C. (2008). Priority setting in health: origins, description and application of the Australian Assessing Cost–Effectiveness initiative. *Expert review of pharmacoeconomics & outcomes research*, *8*(6), 593-617.

Caspersen, C. J., Powell, K. E., & Christenson, G. M. (1985). Physical activity, exercise, and physical fitness: definitions and distinctions for health-related research. *Public health reports*, *100*(2), 126.

Champion, D. J., Ransom, S. M., Lazarus, P., Camilo, F., Bassa, C., Kaspi, V. M., ... & Stappers, B. W. (2008). An eccentric binary millisecond pulsar in the galactic plane. *Science*, *320*(5881), 1309-1312.

Cheng, S. H., Shih, C. C., Yang, Y. K., Chen, K. T., Chang, Y. H., & Yang, Y. C. (2013). Factors associated with premenstrual syndrome—A survey of new female university students. *The Kaohsiung journal of medical sciences*, *29*(2), 100-105.

Cho, S. H., & Kim, J. (2010). Efficacy of acupuncture in management of premenstrual syndrome: a systematic review. *Complementary therapies in medicine*, *18*(2), 104-111.

Cirillo, P. C., Passos, R. B. F., do Nascimento Bevilaqua, M. C., López, J. R. R. A., & Nardi, A. E. (2012). Bipolar disorder and Premenstrual Syndrome or Premenstrual Dysphoric Disorder comorbidity: a systematic review. *Revista brasileira de psiquiatria*, *34*(4), 467-479.

Cobiac, L. J., Vos, T., & Barendregt, J. J. (2009). Cost-effectiveness of interventions to promote physical activity: a modelling study. *PLoS medicine*, *6*(7), e1000110.

Cobiac, L. J., Vos, T., & Barendregt, J. J. (2009). Cost-effectiveness of interventions to promote physical activity: a modelling study. *PLoS medicine*, *6*(7), e1000110.

Cousins, S. O. B., & Tan, M. (2003). Sources of efficacy for walking and climbing stairs among older adults. *Physical & Occupational Therapy in Geriatrics*, *20*(3-4), 51-68.

Cousins, S. O. B., & Tan, M. (2003). Sources of efficacy for walking and climbing stairs among older adults. *Physical & Occupational Therapy in Geriatrics*, *20*(3-4), 51-68.

Cruz, S. Y., Fabian, C., Pagán, I., Ríos, J. L., González, A. M., Betancourt, J., ... & Palacios, C. (2013). Physical activity and its associations with sociodemographic characteristics, dietary patterns, and perceived academic stress in students attending college in Puerto Rico. *Puerto Rico health sciences journal*, *32*(1).

Curl, W. W. (2000). Aging and exercise: are they compatible in women?. *Clinical orthopaedics and related research*, *372*, 151-158.

Curl, W. W. (2000). Aging and exercise: are they compatible in women?. *Clinical orthopaedics and related research*, *372*, 151-158.

Daley, A. (2009). Exercise and premenstrual symptomatology: a comprehensive review. *Journal of Women's Health*, *18*(6), 895-899.

D'Alonzo, K. T., Stevenson, J. S., & Davis, S. E. (2004). Outcomes of a program to enhance exercise self-efficacy and improve fitness in Black and Hispanic college-age women. *Research in nursing & health*, *27*(5), 357-369

Dalziel, K., Segal, L., & Elley, C. R. (2006). Cost utility analysis of physical activity counselling in general practice. *Australian and New Zealand journal of public health*, *30*(1), 57-63.

Danno, K., Colas, A., Terzan, L., & Bordet, M. F. (2013). Homeopathic treatment of premenstrual syndrome: a case series. *Homeopathy*, *102*(1), 59-65.

Dante, G., & Facchinetti, F. (2011). Herbal treatments for alleviating premenstrual symptoms: a systematic review. *Journal of Psychosomatic Obstetrics & Gynecology*, *32*(1), 42-51.

Das, A. K., & Subramaniam, S. R. M. P. (2014). Comparative Analysis of Explosive Strength and Muscular Strength among Government Semi Government and Non-Government School Boys. *Int. J. of Multidisciplinary and Current research.*

Das, A. K., & Subramaniam, S. R. M. P. (2014). Comparative Analysis of Explosive Strength and Muscular Strength among Government Semi Government and Non-Government School Boys. *Int. J. of Multidisciplinary and Current research.*

Davies, S. C., Fowler, T., Watson, J., Livermore, D. M., & Walker, D. (2013). Annual Report of the Chief Medical Officer: infection and the rise of antimicrobial resistance. *The Lancet*, *381*(9878), 1606-1609.

Davydov, D. M., Shapiro, D., & Goldstein, I. B. (2004). Moods in everyday situations: Effects of menstrual cycle, work, and personality. *Journal of psychosomatic research*, *56*(1), 27-33.

De Berardis, D., Serroni, N., Salerno, R. M., & Ferro, F. M. (2007). Treatment of premenstrual dysphoric disorder (PMDD) with a novel formulation of drospirenone and ethinyl estradiol. *Therapeutics and clinical risk management*, *3*(4), 585.

De Bruin, E. D., Hartmann, A., Uebelhart, D., Murer, K., & Zijlstra, W. (2008). Wearable systems for monitoring mobility-related activities in older people: a systematic review. *Clinical rehabilitation*, *22*(10-11), 878-895.

de Vrese, M. (2009). Health benefits of probiotics and prebiotics in women. *Menopause International*, *15*(1), 35-40.

de Zambotti, M., Nicholas, C. L., Colrain, I. M., Trinder, J. A., & Baker, F. C. (2013). Autonomic regulation across phases of the menstrual cycle and sleep stages in women with premenstrual syndrome and healthy controls. *Psychoneuroendocrinology*, *38*(11), 2618-2627.

DeLong, L. L. (2006). *College students' motivation for physical activity* (Doctoral dissertation).

DeLong, L. L. (2006). *COLLEGE STUDENTS'MOTIVATION FOR PHYSICAL ACTIVITY* (Doctoral dissertation, University of Louisiana at Monroe).

Dennerstein, L., Lehert, P., & Heinemann, K. (2012). Epidemiology of premenstrual symptoms and disorders. *Menopause international*, *18*(2), 48-51.

Di Giulio, G., & Reissing, E. D. (2006). Premenstrual dysphoric disorder: prevalence, diagnostic considerations, and controversies. *Journal of Psychosomatic Obstetrics & Gynecology*, *27*(4), 201-210.

Dickerson, L. M., Mazyck, P. J., & Hunter, M. H. (2003). Premenstrual syndrome. *American family physician*, *67*(8), 1743-1752.)

Díez, S. M. U., & Pérez-Fortis, A. (2010). Socio-demographic predictors of health behaviors in Mexican college students. *Health promotion international*, *25*(1), 85-93.

Dishman, R. K., Jackson, A. S., & Bray, M. S. (2010). Validity of processes of change in physical activity among college students in the TIGER study. *Annals of Behavioral Medicine*, *40*(2), 164-175.

Douglas, S. (2002). Premenstrual syndrome. Evidence-based treatment in family practice. *Canadian family physician*, *48*(11), 1789-1797.

Dour, C. A., Horacek, T. M., Schembre, S. M., Lohse, B., Hoerr, S., Kattelmann, K., ... & Greene, G. (2013). Process evaluation of Project WebHealth: a nondieting Web-based intervention for obesity prevention in college students. *Journal of nutrition education and behavior*, *45*(4), 288-295.

Dueñas, J. L., Lete, I., Bermejo, R., Arbat, A., Pérez-Campos, E., Martínez-Salmeán, J., ... & Coll, C. (2011). Prevalence of premenstrual syndrome and premenstrual dysphoric disorder in a representative cohort of Spanish women of fertile age. *European Journal of Obstetrics & Gynecology and Reproductive Biology*, *156*(1), 72-77.

DuMonthier, W. N., Haneline, M. T., & Smith, M. (2009). Survey of health attitudes and behaviors of a chiropractic college population. *Journal of manipulative and physiological therapeutics*, *32*(6), 477-484.

Dunton, G. F., & Schneider, M. (2006). Peer Reviewed: Perceived Barriers to Walking for Physical Activity. *Preventing chronic disease*, *3*(4).

Epperson, C. N., Steiner, M., Hartlage, S. A., Eriksson, E., Schmidt, P. J., Jones, I., & Yonkers, K. A. (2012). Premenstrual dysphoric disorder: evidence for a new category for DSM-5. *American Journal of Psychiatry*, *169*(5), 465-475.

Etats-Unis. Department of health, human services, Centers for disease control, prevention (Etats-Unis), & National center for health statistics (Etats-Unis). (2013). *Healthy people 2010: Final review*. Government Printing Office.

Evenson, K. R., & McGinn, A. P. (2004). Availability of school physical activity facilities to the public in four US communities. *American Journal of Health Promotion*, *18*(3), 243-250.

Fenton, C., Wellington, K., Moen, M. D., & Robinson, D. M. (2007). Drospirenone/Ethinylestradiol 3mg/20µg (24/4 Day Regimen). *Drugs*, *67*(12), 1749-1765.

Ford, O., Lethaby, A., Roberts, H., & Mol, B. W. J. (2009). Progesterone for premenstrual syndrome. *The Cochrane Library*.

Ford, O., Lethaby, A., Roberts, H., & Mol, B. W. J. (2012). Progesterone for premenstrual syndrome. *The Cochrane Library*.

Foster, C., Hillsdon, M., & Thorogood, M. (2004). Environmental perceptions and walking in English adults. *Journal of Epidemiology & Community Health*, *58*(11), 924-928.

Foster, C., Hillsdon, M., Thorogood, M., Kaur, A., & Wedatilake, T. (2005). Interventions for promoting physical activity. *The Cochrane Library*.

Foulds, H. J., Bredin, S. S., & Warburton, D. E. (2011). The effectiveness of community based physical activity interventions with Aboriginal peoples. *Preventive medicine*, *53*(6), 411-416.

Freeman, E. W. (2010). Therapeutic management of premenstrual syndrome. *Expert opinion on pharmacotherapy*, *11*(17), 2879-2889.

Freeman, E. W., Halbreich, U., Grubb, G. S., Rapkin, A. J., Skouby, S. O., Smith, L., ... & Constantine, G. D. (2012). An overview of four studies of a continuous oral contraceptive (levonorgestrel 90 mcg/ethinyl estradiol 20 mcg) on premenstrual dysphoric disorder and premenstrual syndrome. *Contraception*, *85*(5), 437-445.

Global, N. C. D., & World Health Organization. (2011). Global Status Report on Noncommunicable Diseases 2010. *Geneva: World Health Organization*.

Global, N. C. D., & World Health Organization. (2011). Global Status Report on Noncommunicable Diseases 2010. *Geneva: World Health Organization*.

Grubbs, L., & Carter, J. (2002). The relationship of perceived benefits and barriers to reported exercise behaviors in college undergraduates. *Family & Community Health*, *25*(2), 76-84.

Guille, C., Spencer, S., Cavus, I., & Epperson, C. N. (2008). The role of sex steroids in catamenial epilepsy and premenstrual dysphoric disorder: implications for diagnosis and treatment. *Epilepsy & Behavior*, *13*(1), 12-24.

Gyurcsik, N. C., Spink, K. S., Bray, S. R., Chad, K., & Kwan, M. (2006). An ecologically based examination of barriers to physical activity in students from grade seven through first-year university. *Journal of Adolescent Health*, *38*(6), 704-711.

Halbreich, U., Backstrom, T., Eriksson, E., O'brien, S., Calil, H., Ceskova, E., ... & Heuser, I. (2007). Clinical diagnostic criteria for premenstrual syndrome and guidelines for their quantification for research studies. *Gynecological Endocrinology*, *23*(3), 123-130.

Han, H., Lv, X., & Wang, D. (2008). Survey of the dietary status and physical activity level and fitness of college freshmen in Chengdu. *Wei sheng yan jiu= Journal of hygiene research*, *37*(4), 492-494.

Haskell, W. L., Lee, I. M., Pate, R. R., Powell, K. E., Blair, S. N., Franklin, B. A., ... & Bauman, A. (2007). Physical activity and public health. Updated recommendation for adults from the American College of Sports Medicine and the American Heart Association. *Circulation.*

Henneth L Jones and others (1976), "*health science* " 3rd ed. (New York: Harper and Row publishers, p-2.

Heyward, V. H., & Gibson, A. (2014). *Advanced fitness assessment and exercise prescription 7th edition*. Human kinetics.

Indusekhar, R., Usman, S. A. B., & O'Brien, S. (2007). Psychological aspects of premenstrual syndrome. *Best Practice & Research Clinical Obstetrics & Gynaecology*, *21*(2), 207-220.

Jarvis, C. I., Lynch, A. M., & Morin, A. K. (2008). Management strategies for premenstrual syndrome/premenstrual dysphoric disorder. *Annals of Pharmacotherapy*, *42*(7-8), 967-978.

Jerrold S. Green Berg and David Parkman (1986), *physical fitness*, (Englewood cliffs, New Jersey: prentice hall of inc.,), p-3

Jha, S., & Bawa, K. S. (2006). Population growth, human development, and deforestation in biodiversity hotspots. *Conservation Biology*, *20*(3), 906-912.

Jing, Z., Yang, X., Ismail, K. M., Chen, X. Y., & Wu, T. (2009). Chinese herbal medicine for premenstrual syndrome. *The Cochrane Library*.

John Walsh (1968), *the first book of physical fitness,* (London: Man Educational Book Ltd), p-3.

Jones, T. M., Hush, J. M., Dear, B. F., Titov, N., & Dean, C. M. (2014). The efficacy of self-management programmes for increasing physical activity in community-dwelling adults with acquired brain injury (ABI): a systematic review. *Systematic reviews*, *3*(1), 39.

Joshi, J. V., Pandey, S. N., Galvankar, P., & Gogate, J. A. (2010). Prevalence of premenstrual symptoms: Preliminary analysis and brief review of management strategies. *Journal of mid-life health*, *1*(1), 30.

Ju, H., Jones, M., & Mishra, G. D. (2014). Premenstrual syndrome and dysmenorrhea: symptom trajectories over 13 years in young adults. *Maturitas*, *78*(2), 99-105.

Juniper, K. C., Oman, R. F., Hamm, R. M., & Kerby, D. S. (2004). The relationships among constructs in the health belief model and the transtheoretical model among African-American college women for physical activity. *American Journal of Health Promotion*, *18*(5), 354-357.

Kadian, S., & O'Brien, S. (2012). Classification of premenstrual disorders as proposed by the International Society for Premenstrual Disorders. *Menopause international*, *18*(2), 43-47.

Kahn, E. B., Ramsey, L. T., Brownson, R. C., Heath, G. W., Howze, E. H., Powell, K. E., ... & Corso, P. (2002). The effectiveness of interventions to increase physical activity: A systematic review1, 2 1The names and affiliations of the Task Force members are listed in the front of this supplement and at www. thecommunityguide. org. 2Address correspondence and reprint requests to: Peter A. Briss, MD, Community Guide Branch, Centers for Disease Control and Prevention, 4770 Buford Highway, MS-K73, Atlanta, GA 30341. E-mail: PBriss@ cdc. gov. *American journal of preventive medicine*, *22*(4), 73-107.

Kahn, E. B., Ramsey, L. T., Brownson, R. C., Heath, G. W., Howze, E. H., Powell, K. E., ... & Corso, P. (2002). The effectiveness of interventions to increase physical activity: A systematic review1, 2 1The names and affiliations of the Task Force members are listed in the front of this supplement and at www. thecommunityguide. org. 2Address correspondence and reprint requests to: Peter A. Briss, MD, Community Guide Branch, Centers for Disease Control and Prevention, 4770 Buford Highway, MS-K73, Atlanta, GA 30341. E-mail: PBriss@ cdc. gov. *American journal of preventive medicine*, *22*(4), 73-107.

Kahn, E. B., Ramsey, L. T., Brownson, R. C., Heath, G. W., Howze, E. H., Powell, K. E., ... & Corso, P. (2002). The effectiveness of interventions to increase physical activity: A systematic review1, 2 1The names and affiliations of the Task Force members are listed in the front of this supplement and at www. thecommunityguide. org. 2Address correspondence and reprint requests to: Peter A. Briss, MD, Community Guide Branch, Centers for Disease Control and Prevention, 4770 Buford Highway, MS-K73, Atlanta, GA 30341. E-mail: PBriss@ cdc. gov. *American journal of preventive medicine*, *22*(4), 73-107.

Kelly, I. R., Phillips, M. A., Revels, M., & Ujamaa, D. (2010). Contribution of the school environment to physical fitness in children and youth. *Journal of Physical Activity and Health*, *7*(3), 333-342.

Khanday, Z. A., & Akram, M. (2012). Health Status of Marginalized Groups in India. *International Journal of Applied Sociology*, *2*(6), 60-70.

Kilpatrick, M., Hebert, E., & Bartholomew, J. (2005). College students' motivation for physical activity: differentiating men's and women's motives for sport participation and exercise. *Journal of American college health*, *54*(2), 87-94.

Kim, D. R., Czarkowski, K. A., & Epperson, C. N. (2011). The relationship between bipolar disorder, seasonality, and premenstrual symptoms. *Current psychiatry reports*, *13*(6), 500-503.

Kim, H. W., Kwon, M. K., Kim, N. S., & Reame, N. E. (2006). Intake of dietary soy isoflavones in relation to perimenstrual symptoms of Korean women living in the USA. *Nursing & health sciences*, *8*(2), 108-113.

Kim, S. Y., Park, H. J., & Lee, H. (2011). Acupuncture for premenstrual syndrome: a systematic review and meta-analysis of randomised controlled trials. *BJOG: An International Journal of Obstetrics & Gynaecology*, *118*(8), 899-915.

Kimura, N., Fukuwatari, T., Sasaki, R., Hayakawa, F., & Shibata, K. (2003). Vitamin intake in Japanese women college students. *Journal of nutritional science and vitaminology*, *49*(3), 149-155.

Kleinstäuber, M., Witthöft, M., & Hiller, W. (2012). Cognitive-behavioral and pharmacological interventions for premenstrual syndrome or premenstrual dysphoric disorder: a meta-analysis. *Journal of clinical psychology in medical settings*, *19*(3), 308-319.

Koltyn, K. F. (2001). The association between physical activity and quality of life in older women. *Women's Health Issues*, *11*(6), 471-480.

Kondo, T., Kobayashi, I., & Murakami, M. (2006). Effect of exercise on circulating adipokine levels in obese young women. *Endocrine journal*, *53*(2), 189-195.

Kuhl. (2002). Influence of ovarian cycle on the central nervous system. *Therapeutische Umschau*, *59*(4), 175-181.

Kulavic, K., Hultquist, C. N., & McLester, J. R. (2013). A comparison of motivational factors and barriers to physical activity among traditional versus nontraditional college students. *Journal of American College Health*, *61*(2), 60-66.

Kwan, I., & Onwude, J. L. (2009). Premenstrual syndrome. *BMJ clinical evidence*, *2009*.

Lee, I. M., & Skerrett, P. J. (2001). Physical activity and all-cause mortality: what is the dose-response relation?. *Medicine & Science in Sports & Exercise*, *33*(6), S459-S471.

Llorente-Cantarero, F. J., Gil-Campos, M., Benitez-Sillero, J. D., Muñoz-Villanueva, M. C., Túnez, I., & Pérez-Navero, J. L. (2012). Prepubertal children with suitable fitness and physical activity present reduced risk of oxidative stress. *Free Radical Biology and Medicine*, *53*(3), 415-420.

Lo, F. E., Tsai, F. C., Lee, M. B., Tsai, L. T., Lyu, S. Y., & Yang, C. C. (2015). Contextual factors influencing leisure physical activity of urbanized indigenous adolescents. *Journal of the Formosan Medical Association*, *114*(11), 1082-1087.

Lopez, L. M., Kaptein, A. A., & Helmerhorst, F. M. (2012). Oral contraceptives containing drospirenone for premenstrual syndrome. *The Cochrane Library*.

Lubans, D. R., Morgan, P. J., & Tudor-Locke, C. (2009). A systematic review of studies using pedometers to promote physical activity among youth. *Preventive medicine*, *48*(4), 307-315.

Lukes, A. S., McBride, R. J., Herring, A. H., Fried, M., Sherwani, A., & Dell, D. (2011). Improved premenstrual syndrome symptoms after NovaSure endometrial ablation. *Journal of minimally invasive gynecology*, *18*(5), 607-611.

Lustyk, M. K. B., Gerrish, W. G., Shaver, S., & Keys, S. L. (2009). Cognitive-behavioral therapy for premenstrual syndrome and premenstrual dysphoric disorder: a systematic review. *Archives of women's mental health*, *12*(2), 85.

Lzack walron comby Rhoda Thomas Tripp (1976), *the international Thesaurus of Isuatations* (Middle Sex: Pen gain Book), p-6.

Madanat, H., & Merrill, R. M. (2006). Motivational factors and stages of change for physical activity among college students in Amman, Jordan. *Global Health Promotion*, *13*(3), 185.

Maddison, R., Pfaeffli, L., Whittaker, R., Stewart, R., Kerr, A., Jiang, Y., ... & Rawstorn, J. (2015). A mobile phone intervention increases physical activity in people with cardiovascular disease: Results from the HEART randomized controlled trial. *European journal of preventive cardiology*, *22*(6), 701-709.

Makuch, M. Y., Osis, M. J., de Pádua, K. S., & Bahamondes, L. (2013). Understanding and attitudes of Brazilian men with regard to premenstrual syndrome. *International Journal of Gynecology & Obstetrics*, *121*(1), 31-34.

Malinauskas, B. M., Raedeke, T. D., Aeby, V. G., Smith, J. L., & Dallas, M. B. (2006). Dieting practices, weight perceptions, and body composition: a comparison of normal weight, overweight, and obese college females. *Nutrition Journal*, *5*(1), 11.

Malmberg, J. J., Miilunpalo, S. I., Vuori, I. M., Pasanen, M. E., Oja, P., & Haapanen-Niemi, N. A. (2002). A health-related fitness and functional performance test battery for middle-aged and older adults: feasibility and health-related content validity. *Archives of physical medicine and rehabilitation*, *83*(5), 666-677.

Marcus, B. H., Williams, D. M., Dubbert, P. M., Sallis, J. F., King, A. C., Yancey, A. K., ... & Claytor, R. P. (2006). Physical activity intervention studies. *Circulation*, *114*(24), 2739-2752.

Martens, M. P., Buscemi, J., Smith, A. E., & Murphy, J. G. (2012). The short-term efficacy of a brief motivational intervention designed to increase physical activity among college students. *Journal of Physical Activity and Health*, *9*(4), 525-532.

Mashiloane, C. D., Bagratee, J., & Moodley, J. (2002). Awareness of and attitudes toward menopause and hormone replacement therapy in an African community. *International Journal of Gynecology & Obstetrics*, *76*(1), 91-93.

Matlin, M. (2011). *The psychology of women*. Nelson Education.

Matsudo, V., Matsudo, S., Andrade, D., Araujo, T., Andrade, E., de Oliveira, L. C., & Braggion, G. (2002). Promotion of physical activity in a developing country: the Agita Sao Paulo experience. *Public health nutrition*, *5*(1a), 253-261.

Matsumoto, T., Asakura, H., & Hayashi, T. (2013). Biopsychosocial aspects of premenstrual syndrome and premenstrual dysphoric disorder. *Gynecological Endocrinology*, *29*(1), 67-73.

Mayer, D. (2000, September). On the Role of Health in the Economic and Demographic Dynamics of Brazil, 1980–1995. In *conference on Population Dynamics and the Macro Economy organised by Harvard Centre for Population and Development Studies and IFS. September*.

Mazlyn Mena, M., Nagarajah Lee, H. L., Fatimah, A., Norimah, A. K., & Goh, K. L. (2013). Stool patterns of Malaysian adults with functional constipation: association with diet and physical activity. *Malaysian journal of nutrition*, *19*(1).

McDonald, C. M., Widman, L., Abresch, R. T., Walsh, S. A., & Walsh, D. D. (2005). Utility of a step activity monitor for the measurement of daily ambulatory activity in children. *Archives of physical medicine and rehabilitation*, *86*(4), 793-801.

McKercher, C., Schmidt, M. D., Sanderson, K., Dwyer, T., & Venn, A. J. (2012). Physical activity and depressed mood in primary and secondary school-children. *Mental Health and Physical Activity*, *5*(1), 50-56.

Meersseman, W., Cassiman, D., Goossens, W., & Vanderschueren, S. (2008). An unusual cause of syndrome of inappropriate antidiuretic hormone secretion. *Acta Clinica Belgica*, *63*(4), 277-280.

Menopause Guidelines Revision Task Force, A. A. C. E. (2006). American Association of Clinical Endocrinologists medical guidelines for clinical practice for the diagnosis and treatment of menopause. *Endocrine Practice*, *12*(3), 315-337.

Merom, D., Delbaere, K., Cumming, R., Voukelatos, A., Rissel, C., Van der Ploeg, H. P., & Lord, S. R. (2014). Incidental and planned exercise questionnaire for seniors: validity and responsiveness. *Med Sci Sports Exerc*, *46*(5), 947-54.

Merom, D., Phongsavan, P., Chey, T., & Bauman, A. (2006). Long-term changes in leisure time walking, moderate and vigorous exercise: Were they influenced by the national physical activity guidelines?. *Journal of Science and Medicine in Sport*, *9*(3), 199-208.

Modi, M. B., Donga, S. B., & Dei, L. (2012). Clinical evaluation of Ashokarishta, Ashwagandha Churna and Praval Pishti in the management of menopausal syndrome. *Ayu*, *33*(4), 511.

Mokdad, A. H., Marks, J. S., Stroup, D. F., & Gerberding, J. L. (2004). Actual causes of death in the United States, 2000. *Jama*, *291*(10), 1238-1245.

Molnar, B. E., Gortmaker, S. L., Bull, F. C., & Buka, S. L. (2004). Unsafe to play? Neighborhood disorder and lack of safety predict reduced physical activity among urban children and adolescents. *American journal of health promotion*, *18*(5), 378-386.

Morin, S., Biggs, R. W., Sisterson, M. S., Shriver, L., Ellers-Kirk, C., Higginson, D., ... & Dennehy, T. J. (2003). Three cadherin alleles associated with resistance to Bacillus thuringiensis in pink bollworm. *Proceedings of the National Academy of Sciences*, *100*(9), 5004-5009.

Mosallanezhad, Z., Salavati, M., Sotoudeh, G. R., Wikmar, L. N., & Frändin, K. (2014). Walking habits and health-related factors in 75-year-old Iranian women and men. *Archives of gerontology and geriatrics*, *58*(3), 320-326.

Myers, R. S., & Roth, D. L. (1997). Perceived benefits of and barriers to exercise and stage of exercise adoption in young adults. *Health Psychology*, *16*(3), 277.

National Center for Health Statistics (US), & National Center for Health Services Research. (2002). *Health, United States*. US Department of Health, Education, and Welfare, Public Health Service, Health Resources Administration, National Center for Health Statistics.

National Research Council (US). Committee on Physical Activity, Land Use, & Institute of Medicine (US). (2005). *Does the built environment influence physical activity?: examining the evidence* (No. 282). Transportation Research Board.

Nelson, M. C., Lust, K., Story, M., & Ehlinger, E. (2008). Credit card debt, stress and key health risk behaviors among college students. *American Journal of Health Promotion*, *22*(6), 400-406.

Nickelson, J., Alfonso, M. L., McDermott, R. J., Bumpus, E. C., Bryant, C. A., & Baldwin, J. A. (2011). Characteristics of 'tween'participants and non-participants in the VERB™ summer scorecard physical activity promotion program. *Health education research*, *26*(2), 225-238.

O'Brien, P. M. S., Bäckström, T., Brown, C., Dennerstein, L., Endicott, J., Epperson, C. N., ... & Panay, N. (2011). Towards a consensus on diagnostic criteria, measurement and trial design of the premenstrual disorders: the ISPMD Montreal consensus. *Archives of women's mental health*, *14*(1), 13-21.

Osorio, E. A., Raygoza, N. P., & Paloalto, M. L. R. (2013). Barreras percibidas y nivel de actividad física en adultos mayores de Aguascalientes, Ags.: Un estudio transversal. *Enfermería global*, *12*(3), 34-51.

Ossewaarde, L., Van Wingen, G. A., Kooijman, S. C., Bäckström, T., Fernández, G., & Hermans, E. J. (2010). Changes in functioning of mesolimbic incentive processing circuits during the premenstrual phase. *Social cognitive and affective neuroscience*, nsq071.

Pahor, M., Guralnik, J. M., Ambrosius, W. T., Blair, S., Bonds, D. E., Church, T. S., ... & King, A. C. (2014). Effect of structured physical activity on prevention of major mobility disability in older adults: the LIFE study randomized clinical trial. *Jama*, *311*(23), 2387-2396.

Pearson, O. R., Busse, M. E., Van Deursen, R. W. M., & Wiles, C. M. (2004). Quantification of walking mobility in neurological disorders. *Qjm*, *97*(8), 463-475.

Penedo, F. J., & Dahn, J. R. (2005). Exercise and well-being: a review of mental and physical health benefits associated with physical activity. *Current opinion in psychiatry*, *18*(2), 189-193.

Pereira-Vega, A., Sánchez, J. L., Gil, F. L., Maldonado, J. A., Bravo, J. M., Ignacio, J. M., ... & Sánchez, I. (2010). Premenstrual asthma and symptoms related to premenstrual syndrome. *Journal of Asthma*, *47*(8), 835-840.

Phillips, M. M., Goodell, M., Raczynski, J. M., & Philyaw Perez, A. G. (2012). Creating and using index scores in the analysis of school policy implementation and impact. *Journal of School Health*, *82*(6), 253-261.

Physical activity (2014), Retrieved from http://www.betterhealthchannel.com.au/bhcv2/bhcarticles.nsf/pages/Teenage_girls_getting_them_active?open on March 14

Physical activity and health (2014) Retrieved from http://www.cdc.gov/nccdphp/sgr/intro.htm (April 2014)

Physical Activity Guidelines for Americans (2008). U.S.: Department of Health and Human Services. Retrieved February 12, 2013 from:http://www.health.gov/paguidelines/guidelines/default.aspx

Physical activity intervention http://health.mo.gov/data/interventionmica/PhysicalActivity/index%28iframe%29.html (May 2014)

Physical activity level retrieves from http://www.nhlbi.nih.gov/health/health-topics/topics/phys/ (May 2014)

Physical activity level retrieves from http://www.who.int/dietphysicalactivity/factsheet_adults/en/ (May 2014)

Physical activity level retrieves from http://www.who.int/dietphysicalactivity/factsheet_adults/en/ (May 2014)

Physical activity level retrieves from http://www.who.int/topics/physical_activity/en/ (May 2014)

Physical activity retrieves from http://www.nhlbi.nih.gov/health/health-topics/topics/phys/recommend.html (May 2014)

Physical activity retrieves from http://www.who.int/dietphysicalactivity/physical-activity-recommendations-5-17years.pdf Global Recommendations on Physical Activity for Health © World Health Organization 2011

Pimentel, D., Zuniga, R., & Morrison, D. (2005). Update on the environmental and economic costs associated with alien-invasive species in the United States. *Ecological economics*, *52*(3), 273-288.

Pramanik, T., & Pramanik, S. (2001). Physical fitness status among the students of a medical college in Kathmandu, Nepal.

Premenstrual Syndrome retrieves from http://chealth.canoe.ca/condition_info_details.asp?disease_id=227 (May 2014)

Premenstrual Syndrome retrieves from http://chealth.canoe.ca/condition_info_details.asp?channel_id=0&relation_id=0&disease_id=227&page_no=2#Treatment (May 2014)

Premenstrual Syndrome retrieves from http://www.bupa.co.uk/individuals/health-information/directory/p/premenstrual-tension (May 2014)

Premenstrual Syndrome retrieves from http://www.bupa.co.uk/individuals/health-information/directory/p/premenstrual-tension (May 2014)

Premenstrual Syndrome retrieves from http://www.channel4embarrassingillnesses.com/men-in-white-coats/premenstrual-syndrome/introduction/ (May 2014)

Premenstrual Syndrome retrieves from http://www.patient.co.uk/health/premenstrual-syndrome (May 2014)

Public Health Agency of Canada (2011). What Determines Health? – Population Health. Retrieved from, http://www.phac-aspc.gc.ca

Public Health Agency of Canada (2011). What Determines Health? – Population Health. Retrieved from, http://www.phac-aspc.gc.ca

Ramírez-Vélez, R., Tordecilla-Sanders, A., Laverde, D., Hernández-Novoa, J. G., Ríos, M., Rubio, F., ... & Martínez-Torres, J. (2015). The prevalence of barriers for Colombian college students engaging in physical activity. *Nutricion hospitalaria*, *31*(2).

Rapkin, A. (2003). A review of treatment of premenstrual syndrome & premenstrual dysphoric disorder. *Psychoneuroendocrinology*, *28*, 39-53.

Rapkin, A. J., & Akopians, A. L. (2012). Pathophysiology of premenstrual syndrome and premenstrual dysphoric disorder. *Menopause international*, *18*(2), 52-59.

Rapkin, A. J., & Mikacich, J. A. (2008). Premenstrual syndrome and premenstrual dysphoric disorder in adolescents. *Current Opinion in Obstetrics and Gynecology*, *20*(5), 455-463.

Rapkin, A. J., & Winer, S. A. (2008). The pharmacologic management of premenstrual dysphoric disorder. *Expert opinion on pharmacotherapy*, *9*(3), 429-445.

Rapkin, A. J., & Winer, S. A. (2009). Premenstrual syndrome and premenstrual dysphoric disorder: quality of life and burden of illness. *Expert review of pharmacoeconomics & outcomes research*, *9*(2), 157-170.

Rasheed, P., & Al-Sowielem, L. S. (2002). Prevalence and predictors of premenstrual syndrome among college-aged women in Saudi Arabia. *Annals of Saudi medicine*, *23*(6), 381-387.

Reavenall, S., & Blake, H. (2010). Determinants of physical activity participation following traumatic brain injury. *International Journal of Therapy and Rehabilitation*, *17*(7), 360-369.

Reddy, K. N., & Selvaraju, V. (1994). Determinants of health status in India: An empirical investigation. In *The 76th Annual Conference Volume of the Indian Economic Association*.

Ries, A. V., Voorhees, C. C., & Gittelsohn, J. (2010). Environmental barriers and facilitators of physical activity among urban african-american youth. *Children Youth and Environments*, *20*(1), 26-51.

Ries, A. V., Voorhees, C. C., & Gittelsohn, J. (2010). Environmental barriers and facilitators of physical activity among urban African-American youth. *Children Youth and Environments*, *20*(1), 26-51.

Roberts, K. C., Shields, M., de Groh, M., Aziz, A., & Gilbert, J. A. (2012). Overweight and obesity in children and adolescents: results from the 2009 to 2011 Canadian Health Measures Survey. *Health rep*, *23*(3), 37-41.

Rocha Filho, E. A., Lima, J. C., Neto, J. S. P., & Montarroyos, U. (2011). Essential fatty acids for premenstrual syndrome and their effect on prolactin and total cholesterol levels: a randomized, double blind, placebo-controlled study. *Reproductive health*, *8*(1), 2.

Rodgers, W. M., Hall, C. R., Duncan, L. R., Pearson, E., & Milne, M. I. (2010). Becoming a regular exerciser: Examining change in behavioural regulations among exercise initiates. *Psychology of Sport and Exercise*, *11*(5), 378-386.

Rodio, A., & Fattorini, L. (2014). Downhill walking to improve lower limb strength in healthy young adults. *European journal of sport science*, *14*(8), 806-812.

Romans, S., Clarkson, R., Einstein, G., Petrovic, M., & Stewart, D. (2012). Mood and the menstrual cycle: a review of prospective data studies. *Gender medicine*, *9*(5), 361-384.

Rosenbaum, S., Tiedemann, A., Sherrington, C., Curtis, J., & Ward, P. B. (2014). Physical activity interventions for people with mental illness: a systematic review and meta-analysis.

Salamat, S., Ismail, K. M., & O'Brien, S. (2008). Premenstrual syndrome. *Obstetrics, Gynaecology & Reproductive Medicine*, *18*(2), 29-32.

Sallis, J. F., Taylor, W. C., Dowda, M., Freedson, P. S., & Pate, R. R. (2002). Correlates of vigorous physical activity for children in grades 1 through 12: comparing parent-reported and objectively measured physical activity. *Pediatric Exercise Science*, *14*(1), 30-44).

Schatz, D. B., Hsiao, M. C., & Liu, C. Y. (2012). Premenstrual dysphoric disorder in East Asia: a review of the literature. *The International Journal of Psychiatry in Medicine*, *43*(4), 365-380.

Shah, N. R., Jones, J. B., Aperi, J., Shemtov, R., Karne, A., & Borenstein, J. (2008). Selective serotonin reuptake inhibitors for premenstrual syndrome and premenstrual dysphoric disorder: a meta-analysis. *Obstetrics and gynecology*, *111*(5), 1175.

Shah, N. R., Jones, J. B., Aperi, J., Shemtov, R., Karne, A., & Borenstein, J. (2008). Selective serotonin reuptake inhibitors for premenstrual syndrome and premenstrual dysphoric disorder: a meta-analysis. *Obstetrics and gynecology*, *111*(5), 1175.

Sharifi, F., Simbar, M., Mojab, F., & Majd, H. A. (2014). Comparison of the effects of Matricaria chamomila (Chamomile) extract and mefenamic acid on the intensity of premenstrual syndrome. *Complementary therapies in clinical practice*, *20*(1), 81-88.

Sharifirad, G., Charkazi, A., Tashi, M., Shahnazi, H., & Bahador, E. (2011). Physical activity and stages of change among college students. *Health promotion perspectives*, *1*(1), 71.

Shulman, L. P. (2010). Gynecological management of premenstrual symptoms. *Current pain and headache reports*, *14*(5), 367-375.

Shuval, K., Weissblueth, E., Amira, A., Brezis, M., Faridi, Z., Ali, A., & Katz, D. L. (2008). Peer Reviewed: The Role of Culture, Environment, and Religion in the Promotion of Physical Activity Among Arab Israelis. *Preventing chronic disease*, *5*(3).

Sigrist, L. D., Anderson, J. E., & Auld, G. W. (2005). *Senior military officers' educational concerns, motivators and barriers for healthful eating and regular exercise* (No. USARIEM-M05-47). ARMY RESEARCH INST OF ENVIRONMENTAL MEDICINE NATICK MA MILITARY NUTRITION DIV.

Sims, J., Huang, N., Pietsch, J., & Naccarella, L. (2004). The Victorian Active Script Programme: promising signs for general practitioners, population health, and the promotion of physical activity. *British Journal of Sports Medicine*, *38*(1), 19-25.

Slattery, M. L., & Potter, J. D. (2002). Physical activity and colon cancer: confounding or interaction?. *Medicine and science in sports and exercise*, *34*(6), 913-919.

Sleep, I. (2007). Sleep quality and the sleep electroencephalogram in women with severe premenstrual syndrome. *Sleep*, *30*(10).

Smith, L. L., Larkey, L. K., Roe, D. J., Bucho-Gonzalez, J. A., Saboda, K., & Ainsworth, B. E. (2014). Self-Reported physical activity patterns among low-income latina women in Arizona. *Women's Health Issues*, *24*(3), e353-e361.

Sohrabi, N., Kashanian, M., Ghafoori, S. S., & Malakouti, S. K. (2013). Evaluation of the effect of omega-3 fatty acids in the treatment of premenstrual syndrome:"a pilot trial". *Complementary therapies in medicine*, *21*(3), 141-146.

Solomon, C. G., Hu, F. B., Dunaif, A., Rich-Edwards, J. E., Stampfer, M. J., Willett, W. C., ... & Manson, J. E. (2002). Menstrual cycle irregularity and risk for future cardiovascular disease. *The Journal of Clinical Endocrinology & Metabolism*, *87*(5), 2013-2017.

Stamatelopoulos, K. S., Georgiopoulos, G., Papaioannou, T., Lambrinoudaki, I., Kouzoupis, A., Vlachopoulos, C., ... & Sfikakis, P. P. (2012). Can premenstrual syndrome affect arterial stiffness or blood pressure?. *Atherosclerosis*, *224*(1), 170-176.

Steiner, M., Peer, M., Macdougall, M., & Haskett, R. (2011). The premenstrual tension syndrome rating scales: An updated version. *Journal of affective disorders*, *135*(1), 82-88.

Stephenson, J., Bauman, A., Armstrong, T., Smith, B. J., & Bellew, B. (2000). The costs of illness attributable to physical activity in Australian adult population: a preliminary study. *The Australian Sports Commission & the Commonwealth Department of Health and Aged Care*.

Stice, E., Rohde, P., Shaw, H., & Marti, C. N. (2012). Efficacy trial of a selective prevention program targeting both eating disorder symptoms and unhealthy weight gain among female college students. *Journal of consulting and clinical psychology*, *80*(1), 164.

Swartz, A. M., Rote, A. E., Cho, Y. I., Welch, W. A., & Strath, S. J. (2014). Responsiveness of motion sensors to detect change in sedentary and physical activity behaviour. *British journal of sports medicine*, bjsports-2014.

Taguchi, R., Matsubara, S., Yoshimoto, S., Imai, K., Ohkuchi, A., & Kitakoji, H. (2009). Acupuncture for premenstrual dysphoric disorder. *Archives of gynecology and obstetrics*, *280*(6), 877.

Teixeira, A. L. D. S., Oliveira, É. C. M., & Dias, M. R. C. (2013). Relationship between the level of physical activity and premenstrual syndrome incidence. *Revista Brasileira de Ginecologia e Obstetrícia*, *35*(5), 210-214.

Teixeira, P., Silva, M., Vieira, P. N., Palmeira, A. L., & Sardinha, L. B. (2006). A actividade física e o exercício no tratamento da obesidade. *Endocrinologia, Metabolismo & Nutrição*, *15*(1), 1-15.

Terry-McElrath, Y. M., & O'malley, P. M. (2011). Substance use and exercise participation among young adults: Parallel trajectories in a national cohort-sequential study. *Addiction*, *106*(10), 1855-1865.

Thompson Coon, J., Boddy, K., Stein, K., Whear, R., Barton, J., & Depledge, M. H. (2011). Does participating in physical activity in outdoor natural environments have a greater effect on physical and mental wellbeing than physical activity indoors? A systematic review. *Environmental science & technology*, *45*(5), 1761-1772.

Thompson, P. D., Buchner, D., Piña, I. L., Balady, G. J., Williams, M. A., Marcus, B. H., ... & Fletcher, G. F. (2003). Exercise and physical activity in the prevention and treatment of atherosclerotic cardiovascular disease. *Circulation*, *107*(24), 3109-3116.

Tomporowski, P. D., Lambourne, K., & Okumura, M. S. (2011). Physical activity interventions and children's mental function: an introduction and overview. *Preventive medicine*, *52*, S3-S9.

Tremblay, M. S., Warburton, D. E., Janssen, I., Paterson, D. H., Latimer, A. E., Rhodes, R. E., ... & Murumets, K. (2011). New Canadian physical activity guidelines. *Applied Physiology, Nutrition, and Metabolism*, *36*(1), 36-46.

Tudor-Locke, C., McClain, J. J., Hart, T. L., Sisson, S. B., & Washington, T. L. (2009). Expected values for pedometer-determined physical activity in youth. *Research Quarterly for exercise and sport*, *80*(2), 164-174.

Tudor-Locke, C., Pangrazi, R. P., Corbin, C. B., Rutherford, W. J., Vincent, S. D., Raustorp, A., ... & Cuddihy, T. F. (2004). BMI-referenced standards for recommended pedometer-determined steps/day in children. *Preventive medicine*, *38*(6), 857-864.

Usman, S. A. B., Indusekhar, R., & O'Brien, S. (2008). Hormonal management of premenstrual syndrome. *Best Practice & Research Clinical Obstetrics & Gynaecology*, *22*(2), 251-260.

Van der Horst, K., Paw, M. J. C. A., Twisk, J. W., & Van Mechelen, W. (2007). A brief review on correlates of physical activity and sedentariness in youth. *Medicine & Science in Sports & Exercise*, *39*(8), 1241-1250.

Van der Leij, F., Schultz, W. C., van de Wiel, H., & van Leeuwen, J. H. (2009). The premenstrual syndrome. *Nederlands tijdschrift voor geneeskunde*, *154*, A1341-A1341.

Vander Ploeg, K. A., Maximova, K., McGavock, J., Davis, W., & Veugelers, P. (2014). Do school-based physical activity interventions increase or reduce inequalities in health?. *Social Science & Medicine*, *112*, 80-87.

Vaughn, L. M., Jacquez, F., Zhao, J., & Lang, M. (2011). Partnering with students to explore the health needs of an ethnically diverse, low-resource school: An innovative large group assessment approach. *Family & community health*, *34*(1), 72-84.

Vichnin, M., Freeman, E. W., Lin, H., Hillman, J., & Bui, S. (2006). Premenstrual syndrome (PMS) in adolescents: severity and impairment. *Journal of pediatric and adolescent gynecology*, *19*(6), 397-402.

Vickers, K. S., Patten, C. A., Bronars, C., Lane, K., Stevens, S. R., Croghan, I. T., ... & Clark, M. M. (2004). Binge drinking in female college students: The association of physical activity, weight concern, and depressive symptoms. *Journal of American College Health*, *53*(3), 133-140.

Vigod, S. N., Frey, B. N., Soares, C. N., & Steiner, M. (2010). Approach to premenstrual dysphoria for the mental health practitioner. *Psychiatric Clinics of North America*, *33*(2), 257-272.

Wang, Y. J., Hsu, C. C., Yeh, M. L., & Lin, J. G. (2013). Auricular acupressure to improve menstrual pain and menstrual distress and heart rate variability for primary dysmenorrhea in youth with stress. *Evidence-Based Complementary and Alternative Medicine*, *2013*.

Warburton, D. E., Nicol, C. W., & Bredin, S. S. (2006). Health benefits of physical activity: the evidence. *Canadian medical association journal*, *174*(6), 801-809.

Washington, W. D., Banna, K. M., & Gibson, A. L. (2014). Preliminary efficacy of prize-based contingency management to increase activity levels in healthy adults. *Journal of applied behavior analysis*, *47*(2), 231-245.

Watts, P., Phillips, G., Petticrew, M., Hayes, R., Bottomley, C., Yu, G., ... & Lock, K. (2013). Physical activity in deprived communities in London: examining individual and neighbourhood-level factors. *PloS one*, *8*(7), e69472.

Weisz, G., & Knaapen, L. (2009). Diagnosing and treating premenstrual syndrome in five western nations. *Social Science & Medicine*, *68*(8), 1498-1505.

Welk, G. J., Wood, K., & Morss, G. (2003). Parental influences on physical activity in children: An exploration of potential mechanisms. *Pediatric exercise science*, *15*(1), 19-33.

Whelan, A. M., Jurgens, T. M., & Naylor, H. (2009). Herbs, vitamins and minerals in the treatment of premenstrual syndrome: a systematic review. *Can J Clin Pharmacol*, *16*(3), e407-29.

WHO Commission on Social Determinants of Health, & World Health Organization. (2008). *Closing the gap in a generation: health equity through action on the social determinants of health: Commission on Social Determinants of Health final report*. World Health Organization.

WHO, U. (2004). Why are 4 million newborn babies dying each year?. *Lancet*, *364*, 399-401.

Wilkinson, R. G., & Marmot, M. (2003). *Social determinants of health: the solid facts*. World Health Organization.

Wing, R. R., & Hill, J. O. (2001). Successful weight loss maintenance. *Annual review of nutrition*, *21*(1), 323-341.

Wing, R. R., & Hill, J. O. (2001). Successful weight loss maintenance. *Annual review of nutrition*, *21*(1), 323-341.

World Health Organization. (2006). Global strategy on diet, physical activity and health: a framework to monitor and evaluate implementation.

Yan, Z., & Cardinal, B. J. (2013). Perception of physical activity participation of Chinese female graduate students: A case study. *Research quarterly for exercise and sport*, *84*(3), 384-396.

Yang, C. C., Brown, B. B., & Braun, M. T. (2014). From Facebook to cell calls: Layers of electronic intimacy in college students' interpersonal relationships. *New Media & Society*, *16*(1), 5-23.

Zaafrane, F., Faleh, R., Melki, W., Sakouhi, M., & Gaha, L. (2007). An overview of premenstrual syndrome. *Journal de gynecologie, obstetrique et biologie de la reproduction*, *36*(7), 642-652.

Zabinski, M. F., Calfas, K. J., Gehrman, C. A., Wilfley, D. E., & Sallis, J. F. (2001). Effects of a physical activity intervention on body image in university seniors: Project GRAD. *Annals of Behavioral Medicine*, *23*(4), 247-252.

Zhu, X., & Lee, C. (2009). Correlates of walking to school and implications for public policies: survey results from parents of elementary school children in Austin, Texas. *Journal of Public Health Policy*, *30*(1), S177-S202.

Zhu, X., Arch, B., & Lee, C. (2008). Personal, social, and environmental correlates of walking to school behaviors: case study in Austin, Texas. *The Scientific World Journal*, *8*, 859-872.

www.ingramcontent.com/pod-product-compliance
Lightning Source LLC
LaVergne TN
LVHW080047210926
848481LV00011B/3292